Technology-Enabled Blended Learning Experiences for Chemistry Education and Outreach

Technology-Enabled Blended Learning Experiences for Chemistry Education and Outreach

Edited by

Fun Man Fung

Christoph Zimmermann

ELSEVIER

Elsevier
Radarweg 29, PO Box 211, 1000 AE Amsterdam, Netherlands
The Boulevard, Langford Lane, Kidlington, Oxford OX5 1GB, United Kingdom
50 Hampshire Street, 5th Floor, Cambridge, MA 02139, United States

Library of Congress Cataloging-in-Publication Data
A catalog record for this book is available from the Library of Congress

British Library Cataloguing-in-Publication Data
A catalogue record for this book is available from the British Library

ISBN: 978-0-12-822879-1

For information on all Elsevier publications
visit our website at https://www.elsevier.com/books-and-journals

Publisher: Susan Dennis
Acquisitions Editor: Anneka Hess
Editorial Project Manager: Andrea Dulberger
Production Project Manager: Kiruthika Govindaraju
Cover Designer: Alan Studholme

Typeset by SPi Global, India

Contents

SECTION 1 Foundations in technology-enabled blended learning experiences

CHAPTER 1 Theoretical background on technology-enabled learning from an instructional designer's point of view 3

Iman N'hari

CHAPTER 2 Utilizing the power of blended learning through varied presentation styles of lightboard videos 31

Christoph Dominik Zimmermann, Alvita Ardisara, Claire Meiling McColl, Thierry Koscielniak, Etienne Blanc, Xavier Coumoul, and Fun Man Fung

SECTION 2 Curriculum design, implementation, and evaluation, outreach

SECTION 3 Case studies

Contributors

Hafiz Anuar
Department of Chemistry, National University of Singapore, Singapore, Singapore

Alvita Ardisara
Department of Food Science and Technology, National University of Singapore, Singapore, Singapore

Etienne Blanc
University of Paris, UFR Biomedical Sciences; INSERM UMR-S 1124, T3S, Paris, France

Kevin Christopher Boellaard
Department of Chemistry, National University of Singapore, Singapore, Singapore

Michael A. Christiansen
Department of Chemistry and Biochemistry, Utah State University – Uintah Basin Campus, Vernal, UT, United States

Xavier Coumoul
University of Paris, UFR Biomedical Sciences; INSERM UMR-S 1124, T3S, Paris, France

Fun Man Fung
Department of Chemistry, National University of Singapore; Institute for Applied Learning Sciences and Educational Technology (ALSET), NUS, Singapore, Singapore

Jia Yi Han
Department of Chemistry, National University of Singapore, Singapore, Singapore

Garrett Jordan
Galway-Mayo Institute of Technology, Galway, Ireland

Shaphyna Nacqiar Kader
Department of Chemistry, National University of Singapore, Singapore, Singapore

J.L. Kiappes
Department of Chemistry, University College London, London; Corpus Christi College, University of Oxford, Oxford, United Kingdom

Etain Kiely
Galway-Mayo Institute of Technology, Galway, Ireland

Yongbeom Kim
School of Computing, National University of Singapore, Singapore, Singapore

Thierry Koscielniak
Conservatoire National des Arts et Métiers (Le Cnam), DN1 – Direction
Nationale des Usages du Numérique, Paris, France

Yulin Lam
Department of Chemistry, National University of Singapore, Singapore,
Singapore

Elaine Leavy
Galway-Mayo Institute of Technology, Galway, Ireland

Claire Meiling McColl
Department of Sociology, National University of Singapore, Singapore,
Singapore

Iman N'hari
Centre d'Accompagnement à la Pédagogie et SUpport à l'Expérimentation
(CAPSULE), Sorbonne Université, Paris, France

Cormac Quigley
Galway-Mayo Institute of Technology, Galway, Ireland

Max J.H. Tan
Special Programme in Science; Department of Chemistry, National University of
Singapore, Singapore, Singapore

Tag Han Tan
NUS High School of Math and Science, Singapore, Singapore

Teck Kiang Tan
Institute for Applied Learning Sciences and Educational Technology (ALSET),
NUS, Singapore, Singapore

John Yap
Application Architecture and Technology, NUS Information Technology,
Singapore, Singapore

Christoph Dominik Zimmermann
Department of Materials Science and Engineering, National University of
Singapore, Singapore, Singapore

Editors' biography

Fun Man Fung, Ph.D. MSc. BSc., earned his chemistry degrees from National University of Singapore and Technical University of Munich. As a passionate educator, he devotes his time to teach learners how to learn through innovative digital technology. He researches on how gamifications and videography techniques improve learning outcomes. Since 2021, Fun Man is a member of the Editorial Advisory Board at the *Journal of Chemical Education*, ACS Publications. He also served on the International Chemistry Olympiad (IChO) steering Committee and IUPAC Committee for Chemical Education. Fun Man's pedagogical work in Technology-Enabled Learning (2012—2021) is globally recognized. He was conferred a number of accolades for teaching excellence, including the NUS Annual Digital Education Award 2021, D2L award Innovation in Teaching and Learning 2019 and Wharton-QS Reimagine Education Award.

Christoph Dominik Zimmermann, B.Eng., has a background in Materials Science from the National University of Singapore and has been interested in improving education since starting his first educational company in high school in Germany. Working as a research assistant during university, he was exploring technology and gamification to improve tertiary education through the use of 360° videos, lightboards, and boardgames. He is currently working at a consultancy developing AI skilling programs for underserved learners. Christoph is an alumnus of the NUS Overseas College Programme that aims to cultivate and nurture successful entrepreneurs.

Prologue

The COVID-19 global pandemic has put a halt to our usual lives. When this book was conceptualized in 2019, no one would have imagined that the coronavirus was going to overwhelm everyone in extraordinary ways. Nevertheless, this unprecedented crisis offered a renewed opportunity for educators to pioneer teaching methods in the new normal. In this book, we are extremely fortunate to assemble the contributions of scholars, researchers, and innovators from all over the world who are most passionate about improving education in various ways and generous about sharing their unique experiences. We hope that the technology-enabled blended learning experiences illustrated in the follow chapters provide some inspiration for readers who support chemistry education and outreach.

To my mother, who have loved me unconditionally and my family, who stood by me and show that light shall always prevail over darkness.

– Fun Man

To my parents, who have empowered me with their unending support, to my close friend Tom Schamberger, who never fails to open my mind, and to Prof. Fung Fun Man, my mentor and friend.

– Christoph

Preface

Blended learning is a pedagogical approach that combines online learning and classroom learning. This method of teaching and learning is fast becoming a dominant alternative to conventional face-to-face (F2F) lecture and tutorial formats as it promotes flexibility for students to learn at their own pace online and supports student-teacher interactions. As the COVID-19 pandemic shuffles classes to online platforms due to the restrictions, educators find it difficult to implement blended lessons online, especially for those who lacked prior experiences. Our edited book, **Technology-Enabled Blended Learning Experiences for Chemistry Education and Outreach**, serves to inspire educators in thoughtfully integrating technological tools into their blended classroom. Our stories should serve *Science, Technology, Engineering, Art, and Math* (STEAM) educators for the new-looking world moving forward.

This book focuses on three major themes: (1) foundations in technology-enabled blended learning experiences, (2) curriculum design, implementation, evaluation, and outreach, and (3) case studies. In each theme, the authors of each chapter share how they saw opportunities to innovate when faced with teaching challenges, and how they overcame the exigent demands using the right technology adoption. These tools also promote equity and inclusivity in the classroom as they provide the same teaching and learning experiences to educators and students alike regardless of their diverse background. While most of the tools are targeted toward chemistry education, some of the resources are versatile and can be modified to suit the needs of the STEAM educators of nonchemistry disciplines. By introducing the technological tools, we hope that readers will be empowered to try out the tools and integrate them into their curriculum. The tools featured in this book are not exhaustive and serve to slowly bring readers into the world of technology in education, and how technology has an invaluable place in pedagogy, especially in the COVID-19 pandemic era.

Hafiz Anuar and team at the National University of Singapore (NUS) share how they conducted flipped classroom to coach undergraduate students taking the chemistry laboratory module on how to research using a comprehensive online database for the chemical literature—CAS SciFinder. Similarly, Professor Mike Christiansen from the United States, who has 8 years of experience in conducting flipped classroom for freshmen and sophomore undergraduates, share the lessons he had learnt from curating delightful chemistry classes. In addition, Christoph and I also share our experiences on using *lightboard* as an alternative to using conventional whiteboard to teach. In chapter 2, we describe our personal suggestions in filming engaging and cinematic *lightboard* videos.

Readers might also be familiar with the traditional method of delivering didactic lectures when they are pursuing their education. Professor Xavier Coumoul, Professor Etienne Blanc, and team posited that such didactic delivery should be transformed, especially in the context of biochemistry. They proposed a framework called *fragmented course* and shared how they implemented this framework in their biochemistry module to increase interactions between students and teachers, while at the same time allowing students to consolidate their learning through taking part in various engaging activities in class.

Technology plays a pivotal role in education. Therefore, readers will explore some of the available technologies that were used by students in various universities. In chapter 10, Singapore-based John Yap and team introduce to the reader an augmented reality (AR) mobile application (app) to the readers. This AR app was developed for the signature NUS freshman organic chemistry students to help them visualize organic chemical reactions better. With the added gamification element in this AR app, students will be able to interact with the app's interface to elicit the chemical reactions and observe how the molecules undergo changes as the reactions progress. Elsewhere in the United Kingdom, Dr. J.L. Kiappes illustrated how mobile apps further support students' learning in organic chemistry in the reign of visualization, hosting gamified problem-solving questions, and foster collaboration between learners. Kiappes brings his rich teaching experiences from the University of Oxford and University College London in scaffolding the learning processes.

As an extension to organic chemistry, nuclear magnetic resonance (NMR) spectroscopy is an integral tool in laboratories that allows chemists to identify the compounds present in a sample. Tan et al. shared how the use of Mnova, an NMR data processing software, was introduced to undergraduate chemistry laboratory module. Through the software's intuitive interface, students were able to process their raw NMR spectrum independently instead of relying on the lab technician. This self-directed activity helps students to build their critical thinking skills and provides them a head-start in research.

Providing personalized feedback to student is important in their learning. However, this process will be tedious, especially for classes with large student population. Dr. Cormac Quigley and Dr. Etain Kiely overcame this problem by making use of Moodle—a virtual learning environment (VLE) platform—and Microsoft Office to simplify the process. Based in Ireland, Quigley and Kiely shared their experience using the platform to provide personalized feedback to students. With two other colleagues, they also offered another chapter that illustrates the pertinence of learning analytics from VLE that provided insights on learners' behavior, so that educators can design blended learning experiences better to improve students' learning.

Elsewhere in France, Dr. Thierry Koscielniak provided a review on the immersive technologies that were featured in EDUCAUSE Annual conferences since 2016. In this review, Koscielniak provided a brief description of each feedback and how the authors operated. This gives a new perspective on the technologies available for education and a motivation to adopt these technologies as part of their curriculum. Pedagogical methods that enhance students' learning experience are constantly improving, and instructional designers play a major role in such advancement. From her Parisian experience at Sorbonne University, Iman N'hari introduces what it means to be instructional designers and the pivotal roles that they play in pedagogy. N'hari also shared the experiences of the instructional tools that were designed to allow readers to appreciate the important work that such lesser-known occupation has worked.

We conclude by expressing our gratitude to our authors for their supreme support in contributing their work during the tumultuous period in the global health crisis. We register our appreciations to the peer reviewers, whose dedication in providing salient feedback enhanced the quality of this book. We further acknowledge the extensive help and guidance of the Elsevier Books editorial staff, which include Andrea Dulberger, Srinivasan Bhaskaran, Kiruthika Govindaraju, and Anneka Hess. Thank you to all of you! We could not have achieved this without you.

Fun Man Fung
YSEALI Professional Fellow '19

Christoph Dominik Zimmermann
NUS Overseas Colleges Alumnus '18

Acknowledgments

I thank Christoph Zimmermann, my former student, and friend, for working with me in this book. Together, we braved through the era of the COVID-19. I am grateful to have worked with amazing people in Senpai Learn. I am extremely fortunate to have the unwavering support of my thesis advisor, Professor Sam Li Fong Yau, who nurtured my academic growth. I take this opportunity to express my deepest gratitude to my family for being my rock all these years. *To Mom, Mui, Prof. Li, Christoph, and all, thank you so much.*

Fun Man Fung

To my parents, who have empowered me with their unending support, to my close friend Tom Schamberger who never fails to open my mind and to Prof. Fung Fun Man, my mentor and friend.

Christoph

Foundations in technology-enabled blended learning experiences

Theoretical background on technology-enabled learning from an instructional designer's point of view

Iman N'hari

Centre d'Accompagnement à la Pédagogie et SUpport à l'Expérimentation (CAPSULE), Sorbonne Université, Paris, France

Introduction

Unknown: What do you do for a living?
Me: I work at the university, at Sorbonne University.
Unknown: Oww, so you are a teacher, in what subject are you a teacher in?
Me: No, I'm not a teacher, I'm an instructional designer.
Unknown: An instructional what ? What is it ? What does it consist of?

It's the kind of conversation I often have with people I meet at friends' houses, when carpooling… the job of an instructional designer is totally unknown to the general public.

No wonder, because I didn't know this profession myself 10 years ago, before I came across it by chance during my studies at Kingston University in England.

Moreover, the job of an instructional designer can be different depending on the context in which the position is practiced. For instance, in France, we talk of "concepteur pédagogique" literally translated as an "educational designer," any person involved in the creation and mediatization of digital resources; we also talk of an "engineer in information technology and communication for education," any person whose mission is to ensure the coordination, design, production, and development of off-line and online multimedia products and services (websites, front end, intranet) in phase with the development policy of ICTE of the institution for a little over 2 years has emerged the term of "digital education engineer," and distinguishing from the previous one by its specificity to digital, its role is to study and ensure the project management of digital projects to meet the needs of stakeholders and public institutions and promote the evolution of educational practices. All these names demonstrate the extent to which this very recent profession (a little more than 10 years) in Europe is malleable, it evolves with the experience, and it takes the form of the structure in which it is inserted, which gives it a rather unstable character. So you will clearly understand that an educational or pedagogical engineer, which is finally

Technology-Enabled Blended Learning Experiences for Chemistry Education and Outreach
https://doi.org/10.1016/B978-0-12-822879-1.00001-9

the best known and most used designation in Europe, will have different missions depending on whether they work in a university, an engineering school, a training organization, or in the business world.

For my part, the definition that I give to my profession and as I would like it to be described is very close to the definition of an "engineer in information and communication technology for teaching." Indeed, I consider the pedagogy as being at the heart of all issues related to teaching. Tools and technologies are only means to serve it.

In this chapter, I will refer to the term "instructional designer" well known and used in anglophone countries. More generally, what I found interesting in this acronym is the term design. The term "instructional designer" appeared in the 1960s, and it emphasizes a faculty of creation and of invention on the part of the instructional designer more than on the part of his engineering faculty, which contains the idea of a certain scientific technicality.

I have always been fascinated by this plural definition of the profession, which goes hand in hand with the diversity of the people who embody it. If you ask all the people working in this profession, none of them will have the same academic background. This is naturally explained by the fact that no diploma yet existed. For most, educational engineering was a revelation during a university career or professional experience. This was precisely my case.

During my third year of my Bachelor's degree in English Literature and Civilization at Kingston University in London as an Erasmus student (2007), I was fascinated by this teacher who offers French courses in the English language in blended learning, a term totally unknown to me at the time. What method is this? How does it work? I got closer to the French department of the university and sympathized with the head of the department because an idea had just emerged in me, why not dedicate my Master's research thesis to the analysis of this method. So I continued with a Master's degree in French as a Foreign Language and wrote a dissertation entitled "blended learning: the case of English-speaking learners of French at the University of Kingston-London" under the supervision of Anika Falkert from the University of Avignon. This is how I discovered this occupation and how I embarked on this new adventure. What fascinated me a lot during this research thesis is the complexity of this job at the crossroads between Education Sciences, Cognitive Sciences, Human and Social Sciences, Engineering Sciences. I have been greatly influenced by the writings of Marcel Lebrun, Philippe Merieu, Jean Piaget, late, Garrison, Viau, Vygotski, more recently Parmentier, Perraya, Jezegou, and many others.

The importance of the job of instructional designer lies almost essentially in the monitoring and constant constitution of a corpus of articles and books that will accompany him in his search for new practices, feedback from colleagues, and his ability to transpose practices in his context of practice and according to his own affinities. Digital is not the answer to everything when a teacher comes to you with a desire to revitalize his or her course.

The answer that the instructional designer will give will depend a lot on the teacher himself, his ease or not with the tools, the time he wants to devote to it, the primary reasons that lead him to shake up his habits, and his confidence in instructional designing. The instructional designer is very often poorly identified in his structure. Some consider it as a job supporting the use of certain tools, others see a real added value in his role of accompanying the implementation of pedagogical systems, and for some, this job is of no use at all.

Why do we need an instructional designer?

Because we were able to briefly underline in the previous chapter, an instructional designer designs learning systems and assists actors, especially teachers, if it is a university context, in the implementation and deployment of its systems.

A teacher-researcher is valued more for his or her research than for his or her teaching. This leads to cases of teachers who have not questioned their teaching for more than 20 years. Most of them have a desire for "modernization" and have very quickly understood that being an expert in the content of one's course is not enough to master and develop teaching.

Some teachers go even further and look at the notion of learning to learn because this is the primary role of a teacher, namely his or her ability to question how my students learn. It is in this way that the teacher questions what effective learning is and how to make my course effective; it is precisely in this questioning that the instructional designer intervenes. Bloom in his taxonomy in 1957 listed all the action verbs corresponding to the multiple ways of learning: describe, observe, deduce, produce, comment, compare, analyze…. To all these action verbs correspond multiple strategies and resources that should be carefully chosen to make the learner active and receptive.

The instructional designer in his mastery of the different learning strategies will be able to coconstruct with the teacher a teaching in accordance with a coherent pedagogical alignment. The instructional designer will be able to rely on a method in pedagogical project management very well known as the ADDIE method (Analysis, Design, Development, Implementation, Evaluation) (we will come back in detail to what these acronyms contain in another chapter).

Often compared as a "craftsman of knowledge," an instructional designer knows where to draw its resources from several pedagogical currents: cognitivism for the encoding of knowledge, humanism for the learning climate, constructivism for the construction of knowledge, socioconstructivism in coconstruction, and more recently, connectivity for networked learning.

As introduced by G. Miller and J. Bruner in 1956, in the cognitivist approach, there is the idea, contrary to the behaviorist current that preceded it, which defines the teacher as the holder of knowledge, that the teacher plays a central role in helping students acquire and construct new knowledge. The learner is then more active and autonomous in the organization of his knowledge and the links he could make with events in his own personal experience.

Piaget in 1975 went further by developing the notion of constructivism. He then recalled that each learner has knowledge and skills that enable him to build other knowledge and other skills to solve the problems posed by his environment.

"The learner steers his learning, by building his knowledge, he builds himself and by building himself, he acquires his knowledge, this is the very basis of constructivism." Learning is, therefore, seen as a process of construction and not as a process of simple acquisition. Teaching activities are, thus, seen as a means of building knowledge and not as activities for the transmission of knowledge.

Ten years later, Vygotsky enriched the notion of constructivism by adding the prefix "socio," which requires another dimension, namely the enrichment of knowledge through interaction. The learner is indeed at the center of the learning process, he is autonomous, and builds his knowledge from his interactions and sociocultural environment. The teacher's role is, therefore, to encourage interaction between students.

More recently, in 2005, Siemens and Downes introduced the notion of connectivity to refer to learning in relation to new technologies and the possibility of "learning through all the interactions allowed by networks." The learner becomes the master of his knowledge, of the construction of his knowledge that he must confront through collaboration and the search for resources.

It is very important for an instructional designer to sensitize teachers to a socio-constructivist or even connectivist approach and to encourage collaborative work. In parallel with these currents, it is also interesting to question the teacher on his teaching approach, which can be classified into four categories: inductive, deductive, analogical, and dialectical. Very often, teachers use a deductive approach because it corresponds to the way they themselves have learned, and it is also the one that does not question the whole construction of teaching.

The instructional designer questions, analyzes, challenges, and transforms while relying on these theories of learning to support and substantiate his arguments. They create material by also relying on methods and tools that enable them to set up distance learning or so-called "blended learning" approach.

Theoretical background
The ADDIE model (Analysis, Design, Development, Implementation, Evaluation)

The ADDIE model is a project management method applied to learning systems and used by instructional designer in the conduct of pedagogical projects. A project born from a need that has been identified by the teacher who remains the expert in his classroom. Very often, a teacher calls upon a instructional designer to improve his teaching to make it more attractive, more interactive, more dynamic, and more collaborative. The teacher is subject to a university teaching program, a pedagogical model and evaluation methods defined by the institution to guarantee the value of the diploma. On the other hand, the teaching methods are left to the choice of the teacher, who remains the expert in his or her subject and content. It comprises five cyclical phases:

Analysis: Analysis of the context, training needs, definition of the public concerned, definition of the theme and its objectives, identification of stakeholders (internal resources or resources to be recruited if necessary), expectations, constraints, etc.

Upstream of this project scoping phase, it is interesting to provide the various project stakeholders with a scoping document containing a certain number of relevant questions to refocus the project as a whole (Example of questions: make a state-of-the-art of the existing, what will be the added value of my project? what is the target audience? Is it a specialist public, is it open to the general public? What is the need for this teaching? What are the overall objectives? How can they be achieved? Will there be collaborative work (group projects…), what technologies will be used? What is the cost (human, financial)?

Expected deliverable: design brief and requirement specifications.

Design: Definition of content types and learning scenario and definition of objectives and subobjectives with reference to Bloom's taxonomy and answering the formula "By the end of this course, you will be able to…," it is, thus, necessary to determine objectives that must be SMART; that is, meaningful, measurable, attainable, realistic, and within a given time, definitions of learning activities and evaluations, description of the detailed plan specifying the pedagogical resources to be designed and in what format (video, graphics…), and definition of the pedagogical scenario taking into account the chosen learning platform (for the construction of the syllabus), taking into account the accessibility aspect and the collection of data to be used afterward and the ethical framework defined.

A zoom on the objectives: The learning objectives are on three levels: general, pedagogical, and specific as Jean Loisier points out in the Guide to e-learning practices.

The general objectives focus on mastering a field of knowledge related to a social, professional, or personal role. The pedagogical objectives are used to determine the learning content required to achieve the general objectives. The specific objectives that lead to the choice of pedagogical activities are used to evaluate the student's ability to have acquired a skill or not. It is at the level of the specific objectives that most of the training planning (design) is done. This stage in online teaching is crucial because it determines the type of activities that are possible at a distance (laboratory experiments, possible in a classroom setting and difficult to reproduce at a distance).

These objectives should make it possible to target three main categories of knowledge: knowledge, know-how, and interpersonal skills. Although knowledge is available in excess on the Internet, the teacher's role will be to help the student to sort, select, synthesize, and compare all the information.

The knowledge-being will be of a social and relational nature (working in groups and speaking in public). Know-how refers to production activity in a given context.

The notion of competence that results from this is indeed a set of knowledge, know-how, and the ability to be mobilized in a context. The teacher must, therefore, construct activities that enable the development of skills while contextualizing it. Group projects encourage this approach.

Expected deliverable: detailed synopsis

- **Development**: content writing, creation of scenarios, production of the course, media coverage, and validation of scientific content by stakeholders.

  Expected deliverable: scripts and storyboards

- **Implementation**: dissemination of the course, implementation of organizational and technologicalsupport methods, and coaching activities.
- **Evaluation**: evaluate the proposed course using measurement tools (survey, poll, and interviews) to collect Learning Analytics.

Because the ADDIE model is efficient but seen as a bit rigid, it is possible in the manner of agile methods to carry out iterative developments by designing prototypes or pilot projects and find evaluation at every step of the process to check, control, and validate.

However, other models derived from the ADDIE model exist with their own specificities. We can cite MISA Method for Engineering and Learning System, the 4C/ID model, the agile SAM model, or Design Thinking, which is seen rather as a cocreative approach associating analytical and intuitive thinking.

Very often, to save time, it is interesting for the instructional designer to dimension with the teacher the project and the investment time of each one to use the best possible method.

The ADDIE method was used in the creation of a MOOC in public health, which will be presented in the "Application" section of this chapter.

Training facilities

With the advent of new technologies and the massification of education, the university must constantly renew and modernize itself to meet the demands of an ever-changing society, which places lifelong learning as an investment and knowledge as a guarantee of competitiveness. Consequently, for more than 20 years now, universities have been setting up services whose main missions are the development of new technologies in education.

The choice of a training system depends very much on university policy. Generally speaking, we distinguish four types of course delivery modalities: unimodality (asynchronous distance training or self-training aimed at individual distance learning), bimodality (class groups are made up of learners simultaneously in presence and at a distance by videoconfence), comodality (simultaneous mix of a physical class, a synchronous virtual class, and an asynchronous virtual class), and multimodality. The multimodal course is often associated with continuing education for which training needs and expectations vary considerably from one project to another (blended learning, MOOC...).

Blended learning approach

These new technologies such as online and distance learning platforms are becoming essential tools for providing teachers and students with new ways of teaching and learning. The said "blended" devices are then considered as an effective means of learning because it is defined as "Any training device (courses, in-service training) based on a digital environment (e-learning platform) and offering students resources to be used or activities to be carried out at a distance (outside classrooms) and in person (in classrooms)" [1].

The mistake, however, would be to consider blended learning as an addition of distance and face-to-face activities without really taking into account the system as a whole and to consider distance as an absence. A blended system requires rethinking one's teaching and structuring it in such a way as to identify the activities that can be conducted in presence and those that can be conducted at a distance.

It is, therefore, important for a teacher to propose a syllabus on the learning platform containing the following headings: identification of the course, that is, the title of the course and the number of ECTS credits. It may be interesting to specify the prerequisites of the course, teacher information, that is, name, brief biography or bibliography, office location, contact information, and hours of reception, course description, and learning objectives, that is, what is the course intended to teach the students? what should they be capable of at the end of the course? program and timetable, course materials, instructions for exercises or work, individual or group, arrangements for assessment of learning, that is, what will the final examination consist of, when will feedback on the work be given to students, how will the final grade be calculated, and what are the assessment criteria? course operating principles: it can be very useful to make it clear to students how the course will be run by explicitly stating their expectations in terms of participation. This syllabus is a bipartite contract between the teacher and the student and is a guarantee of success.

The temporality of the resources (self-paced...) must also be indicated to give the student an indication of the time spent reading a document, analyzing a video, and answering an exercise, for example.

One of the main challenges of blended learning is to be able to maintain a balance between distance and face-to-face teaching and, thus, ensure that students are constantly motivated over time. In a conference held at CAPSULE: Center of Pedagogy and Support to Experimentation of Sorbonne Université in 2017 (where I work), Laurent Cosnefroy, Professor in Sciences de l'Education at Normal Sup in Lyon, and Louise Menard, Professor in the Didactic Department at UQAM, stress the importance of motivation support because motivation is not innate, especially in front of complexity. If the student's objective is very often to succeed in his semester, which is too distant as an objective, the teacher's role will be to set a list of intermediate objectives that translates into a process of instructional designing such as the granularization of learning tasks. These tasks must also be set within a time frame. Thus the notion of personal efficiency and satisfaction with the task accomplished appears.

According to COMPETICE [2], a steering tool for ICTE and open and distance learning policies, there are five types of blended learning scenarios with a more or less important part of distancing.

– Enriched presential: 100% presential (diversity of media resources), enhanced presential: upstream/downstream with resources to be consulted remotely and remediation activities in the classroom, lean attendance: 50% of face-to-face and TD activities replaced by collaborative activities and self-study modules, and reduced presential: 10%–20% of the activities in presence and nonexistent presence: 100% online.

Between 2009 and 2012, DG EAC (Directorate General for Education and Culture) has launched a research program called Hy-SUP [1] to study hybrid devices in the European educational landscape and understand their effects on learning and teaching practices. Using a corpus of 200 questions, the first study identified six types of hybrid devices: "the stage," "the screen," "the lodging," "the crew," "the metro," and "the ecosystem."

The stage is a device based on face-to-face teaching and containing mediated resources. The screen focuses on the teaching and its content and the student's freedom to consult the resources. The gite is a device focusing on teaching, its content integrating resources from external speakers. The platform is then used for the management of homework and assignments, and the teacher intervenes very little in the organization of the student's tasks remotely. The crew is a device more focused on learning by providing students with learning support tools, communication support tools (forum), and multimedia resources and states objectives. The metro is a device that emphasizes the active participation of face-to-face and distance learning students in student collaboration. At a distance, students are guided step-by-step through the project. The platform serves both as a collaborative space (forum) and as a space for rendering work (peer review). As for the ecosystem, it brings together all the characteristics seen earlier, namely the active participation of students, both in person and at a distance, the frequent and diversified use of technological tools, the provision and encouragement of multimedia document production, interaction between peers, and the opening of the system to external resources and actors.

The blended approach admits three steps: before, during, and after class. The blended learning as well as distance learning are becoming a future issue, especially in the current health context of COVID-19. Numerous studies, including those of the HY-SUP project, demonstrate the positive impacts of hybridization on teaching, which can be classified into several categories: motivation, learning resources mobilized, skills developed, interactions, student activities, and student work. What essentially emerges is a greater number of students active in the course, more varied and contextualized resources, the ability of students to organize their learning, to conduct documentary research, to organize more group work, and to better evaluate student progress and improve their productions.

When a teacher wishes to develop a hybrid course, he or she should bear in mind the 14 components developed by European HY-SUP research, which are as follows:

In my system, I have implemented the following: (1) active participation in the presence, (2) active participation at a distance, (3) learning support tools (4) management, communication, and interaction tools, (5) multimedia resources, (6) work, (7) synchronous communication and collaboration tools, (8) commenting and annotating online documents, (9) reflective activities, (10) methodological support, (11) student support, (12) metacognitive support, (13) freedom of methodological choices, and (14) use of external resources.

The teacher must be able to balance time and workload between distance and presence and act as a facilitator. He must provide clear instructions (response times, resources, spaces, configuration of repository spaces, communication spaces, set up formative activities, and manifest himself during distance time through the use of virtual classrooms, time for exchanges, and answers to students' questions). The face-to-face periods are then intended for reflective activities.

MOOC (Massive Open Online Courses)

The MOOC is an open and free training facility capable of accommodating a very large number of participants. This solution has emerged in response to the massification of teaching.

A MOOC is generally based on the transmission of knowledge offering guidance through an exchange on a forum. A MOOC can be certified. The contents are most often progressive, and the evaluation is based on MCQs or peer-reviewed work.

MOOCs in France are a phenomenon that grew in 2013 with the creation of the FUN MOOC platform prescribed in the law of 22 July 2013, a "law of openness (to the socio-economic environment, to the international arena, to all forms of teaching and research) and a law of transformation."

A MOOC is definitely not a way to replace the lecture. It is not a simple video course either. MOOCs deal with diverse and varied themes according to a sequential and chronological plan that favors the learner's progression. It is built with granularized sequencing and defined pedagogical objectives. Multimedia resources do not exist on their own. They are accompanied by a variety of activities offered by the platform. The strength of an MOOC lies in the collaborative exchange tools and the large community of learners who can come together.

The tutor(s) and forum moderator(s) who are usually the authors of the MOOC take pleasure in actively participating in the questions asked by the learners and showing their interest in the MOOC subject. The MOOC is open. It is, therefore, a question of giving access to resources to as many people as possible by trying to popularize them and, if necessary, to provide in-depth resources that will enable us to go further.

Contrary to the American approach to MOOC, which seems to respond in part to the problem of access to education for all, in France, an MOOC does not replace either teaching or a teacher. MOOC resources are used and reused by the authors as a kind of showcase for their teaching.

It is more a question of notoriety. Some MOOCs in France are certified, most are not. Other initiatives in MOOC called SPOC (private online courses) have been successful with universities because the investment was targeted at student audiences and addressed the problem of massification in Bachelor's degrees. In an MOOC, there is the idea of training freely among the diversity of themes proposed.

What prevents the MOOC from flourishing is what has long been decried as a motivational problem related to this teaching over several weeks and engaging and without synchronous feedback.

The MOOC in public health, which will be presented in the application section of this chapter, will allow you to appreciate the complexity of the device and all the human resources involved.

Designing a blended learning module

The pedagogical scenario plays a central role in the design of a hybrid device. Scriptwriting means above all taking into account the context, defining learning resources and activities, thinking about evaluations, interactions, feedback modes, support modes, and workload. The instructional designer plays a key role in the implementation of the system by questioning the teacher about his practices and what he wants to achieve.

Very often, the multimedia resources created are a heavy investment at the outset, and the design may seem time-consuming for a long-term return on investment that is far more than advantageous for the teacher and the institution.

An instructional designer must support the teacher in the implementation of an active and collaborative approach and explain the change in the teacher's stance toward the class.

The instructional designer is not so much concerned with the content as with the coherent and logical sequence of learning activities. He or she focuses on the types of knowledge and their levels of complexity. His added value lies in his ability to question the teacher on the targeted learning areas: are we in a cognitive domain (knowledge) with Bloom's taxonomy (understand, analyze, synthesize, evaluate, create), are we in the affective domain (being) of Krathwhol (getting involved, committing, and organizing one's values) or are we in the domain of Jewett's know-how (imitate, execute, adapt, compose, and vary). In the construction of the scenario, the instructional designer is not only involved in the organization of the contents but also in the way of finding effective strategies to make learning active and support motivation.

An effective pedagogical scenario is one that integrates a diversity of pedagogical strategies and a multiplication of resources to personalize the courses as much as possible. Scenarization allows the skeleton or backbone of a course to be built. A learning scenario must be able to take into account the context of the learning situation (constraints, resources, and needs), the types of knowledge mobilized and the objectives that underlie them, and the pedagogical activities, productions, and work to be conducted. Several scenarios can then be envisaged [3] in addition to the learner support scenario, namely constrained scenario: set of activities to be performed in a prescribed order; free scenario: set of activities to be performed in the order of one's choice; parallel scenario: set of activities to be conducted concomitantly; subunit scenario: activities to be conducted in subunits; scenario with reflexive feedback: reflection on one or more activities before continuing; and personalized scenario: activities to be performed as one of a set of activities.

For the student, according to the choice of scenarios defined by the pedagogical team, the student will be able to choose the most favorable place and time for learning his or her work rhythm. Very often at the university, teaching is organized in large blocks of multimedia documents intended for large groups (bachelor's degree) leaving little room for a collaborative space for expression.

To promote spaces for exchange and avoid the feeling of isolation, the teacher should aim at learning scenarios that encourage interaction. These may be synchronous or asynchronous interactions. Scheduling synchronous (virtual classroom and videoconferencing) and asynchronous (forum and group work) moments of exchange creates a sense of belonging for the learner. Asynchronous moment must be the rule when building a distant course. Synchronous moment must be used to answer students' questions. Teachers must understand that holding a conference online as it was in classroom can be tedious and exhausting.

Evaluation of learning/evaluation of devices

The Teaching and Learning Support Service (TLS) of the University of Ottawa has made available to the educational community an evaluation grid for hybrid systems [4]. This relatively exhaustive grid can be used as a basis by teachers and educational engineers. The grid has four main headings: course design, learner support and resources, use of technology, and course organization and content presentation.

An example of this grid will be displayed in the application chapter to the green chemistry project.

Beyond the evaluation of the system as a whole, let us look at the evaluation of learning as a key element of a successful pedagogical alignment. Evaluation, whether formative or summative, is a good indicator of the learner's progress and makes it possible to target the elements acquired and those that require further study. Whatever form this evaluation takes (grading, percent, etc.), because the question of grading could be raised, the evaluation is an assessment; it marks the end and the beginning of a reflexive act, a learning situation.

Any evaluation situation must be contextualized; otherwise, it makes little sense. As Leroux explains in 2018, evaluation has four intentions: diagnostic, that is, to identify what has been learned and to direct toward relevant learning; formative, that is, to situate the level of mastery of learning, to adjust the approach, to provide feedback, and to support self-evaluation; summative, that is, to gauge the degree of acquisition or the quality of a performance by a result; and certificative, that is, certify the degree of mastery of a stage or a course.

Within the framework of a hybrid system, the instructional designer organizes the evaluation activities within the framework of the pedagogical scenario and with the help of the available tools.

These evaluation activities can be conducted synchronously, asynchronously, individually, or collaboratively. A variety and frequency of evaluation activities that do not necessarily result in a grading is an ideal to be achieved to give reliable indicators of progress to students and to cover the field of expected competencies.

Very often in a hybrid system, the teacher relays diagnostic and formative evaluation activities at a distance. He or she is present through a forum or a moment of synchronous exchange to answer questions and persistent misunderstandings. In a summative evaluation, the students are not allowed to solicit their peers or to cheat.

Leroux [5, p. 12] lists all the assessment tasks and the tools for collecting traces:

Tasks	Learner assessment tools	Instructor assessment tools
Critical analysis, electronic voting, demonstration, demonstration, electronic voting, electronic voting, project, debate, interview, teamwork, exercises, games, role play, simulation, publication, publication, survey, role play, reporting, survey, survey research.	Text document, multimedia production, portfolio, mind map, database, blog article, poster, collaborative text, wiki, glossary, etc.	Test, evaluation grid, and list

Even if the approach of pedagogical models based on socioconstructivism favors an approach based on error, remediation, and collaborative learning construction and invites the teacher to a role of accompaniment rather than an examining role, the straitjacket of the degree and the curricular approaches of universities still remain very much ingrained and necessarily influence assessment methods.

Tools of the training devices

Eighty-six percent, a figure announced by Thierry Koscielniak in 2016, of French universities use the Moodle platform to provide online and distance learning.

For most instructional designers, the media coverage of content is done on the Moodle platform. Hence the dual pedagogical and technical expertise of our profession often leads to confusion and confines us to a skill more technical than pedagogical.

Training teachers to use the tool is often a way for instructional designer to add to the first technical level a pedagogical layer inseparable from the tool. Moodle, similar to most LMSs, is an organized platform allowing to build pedagogical paths around resources and learner-centered activities. Moodle perfectly integrates the model around a socioconstructivist learning. EdX (a nonprofit platform linked to MIT and Harvard), with the exception of its nonopen source character, is similar to Moodle but targeted for MOOC device courses.

An online and distance learning platform becomes for an institution the place of exchange par excellence in an unlimited space-time because I can connect from where I want and at the time I want.

When the instructional designer builds the pedagogical scenario of a course, the tool must not be a brake in the access to resources and contents, which take precedence over the use of technologies. Nevertheless, if the objective is to create distance learning, the technical constraints must be taken into account (even if solutions are always possible).

Teachers have increasingly ambitious projects involving multiple resources and more complex structures. This structuring must not deteriorate the design, navigability, and cognitive overload of the learner. It is, therefore, important to provide a visual signature by working on a graphic charter to harmonize the whole.

For learning to be an engaging, motivating, and active experience, the tool must be used wisely. In this regard, Mayer [6] identifies some principles for publicizing distance learning such as remove all nonessential information; highlight important information; avoid unnecessary repetition across multiple media; divide content into units; and provide key information at the beginning of learning. The visual support marks a form of cognitive imprint and participates favorably in the learner's learning experience. For example, Moodle offers a varied course format that can be adjusted to several types of scripted courses (more familiar thematic format, weekly format, informal, reduced sections, and view by image) and a variety of activities and resources for the mediatization of distance courses: discussion forum, questionnaire, survey tool, evaluation, wiki, work deposit, and others.

The production of multimedia resources comes into play during the development phase of the ADDIE model. Although supported in the production and editing of the videos by a team of multimedia systems engineers, the fact remains that the storyboard must be clear to record a clear discourse and the surrounding elements that will support the discourse. The recording can take place in the studio or outside. Graphic elements can be added to the protagonist's speech always with the aim of supporting understanding. Before starting on the creation of educational videos, the instructional designer takes the time to ask the following questions: Why the video? What does it add to the classroom discourse? What is its usefulness for my students and my teaching? How much time can I devote to making it?

Very often, video is perceived as a way for the teacher to enhance the value of face-to-face collaborative time, to free himself from long hours spent in lectures to promote activities that make the learner active, and, thus, encourage more interaction and motivation. It supports and enhances learning before, during, and after the course. It can have several objectives: to be used as a summary, to answer the last question of the course or as an introduction to the next topic, make more explicit and visual the link between course content and real-life applications, motivate and engage students, to establish an affective relationship with the students, thus promoting a more constructive and effective exchange in class, presenting, in science, procedures that are too dangerous, too complex, and delicate.

Be careful, video is only a resource in a thoughtful and scripted pedagogical path, whatever the pedagogical modality chosen: reverse class and blended learning. Thus it is necessary to understand that, if we have a precise objective to reach, making a pedagogical video cannot be improvized! Here are some steps and advices that are the fruit of my experience. You have to consider three times in the creation of a video:

Before recording: define the target audience, the appropriate language, and the learning objectives; prepare the images to be commented on and the animations on the screen; define the graphic charter, the authors, and credits; and finally, format it in the most detailed storyboard format possible. The speech must be divided into learning units to make the text coherent. The following rule should apply: One video = one learning objective = 5–10 min.

During the recording: the diction must be clear and irreproachable, and the use of a prompter is possible. The teacher should give room for spontaneity and play with the camera to avoid giving a tense look. The subject should be limited to about 5 min to be effective.

After the recording: the editing will be done by an audiovisual team. The validation of the content by the teacher is necessary. This video will be uploaded on the online teaching platform and available to students at any time. In addition to the segmentation of the information, it is important to choose the desired video format. Depending on the objective, the video can be:

– A screencast (screen + voice), an assonorized slideshow with web cam, an assonorized slideshow with web cam to also film the whiteboard.

Application
Case study number 1: MOOC

In January 2018, UPMC and Paris Sorbonne Sorbonne Université merged to become Sorbonne Université divided into three faculties: Medicine, Sciences, and Humanities.

The center for pedagogical support and experimentation, CAPSULE, belongs to the Faculty of Sciences. It has the potential to enhance, support, and evaluate learning and teaching. It also has human resources such as instructional designers, a video production team, IT…. Although the political and strategical view of the Faculty of Sciences is more oriented on the creation and design of SPOCs, some MOOCs were initiated based on the personal motivation of some teachers.

MOOC at the faculty of medicine

The faculty of medicine has it is own video production team and a graphic artist but no instructional designers. To support those initiatives and help teachers to implement a methodology on the designing process of MOOCs, crossed innovative experiences are encouraged between the three faculties, which among them are the exchange of skills, particularly those of instructional designing.

The following will present the way to manage an MOOC on public health subject with the particular example of the MOOC on obesity.

Experience feedbacks

An MOOC on public health: What for?

This MOOC is the first one in France dealing with "the care of the obese person." Obesity is a serious disease requiring a long way process of treatments and, thus, for life. It is necessary to have a multidisciplinary team of doctors to take care of a person obese. Nowadays, obesity is not considered as a disease and, thus, not reimbursed by the social security. It was important that this MOOC emphasizes on this disease that affects 600 million of adults worldwide. One of the main motivations of the teachers comes from the ambition to inform the population on the constant growth of this disease and then integrating this project in association with the patients and their families and give tools to doctors, dieticians, surgeons and psychologists, and sports coach to treat those patients and give them the right information. The main goal was to inform report and communicate around this disease and build a community of experts. No verified certification was available in this MOOC and that was an initial choice from the team.

A MOOC on public health: The feasibility study

The MOOC "the care of the obese person" is the first MOOC produced by Sorbonne Université team consisting of an instructional designer, a graphist, two videographers and two doctors/nutrionists, and the project leaders implemented in the edX platform. Patients from the association "Vers un nouveau regard" also participated to the MOOC.

Designing the MOOC

The MOOC was created according to the ADDIE model. The MOOC was divided into 5 weeks. A week is made of four to seven sections. Each section is a video of 5–10 min. Videos are illustrated with tremendous graphics created from scratch by the graphics artist of the team.

The doctors define each objectives and pedagogical activities, thanks to a synopsis. The main idea was to keep in mind that they need to tell a story, the story of a patient: from the first contact with the doctor to the ultimate surgery. However, the most important is how the medical team will take care of the patients all along his life.

Some of the videos shooting took place in the hospital in a real-life situation and with a patient to emphasize on the different aspects of the multidisplinary care of the obese person (medical equipment, oversize beds, doors, chairs weight balances…).

Six months have to be dedicated to those two previous phases. Most of the time, teachers and doctors are not prepared to spend that amount of time, and at that time, they realize how difficult could be to build an MOOC. Advertising the MOOC is a part that comes at the 6 last months of the project. This is one of the hardest but at the same time exciting part of the MOOC. We choose to hold a press conference in front of experts and journalists and to use social media to relay the launch of the MOOC with the help of well-known YouTubers.

Throughout the project, the difficult part was to keep deadlines clear in mind and to be aware that they can be adjusted at any time. In addition, the financial costs of human resources (Sorbonne team, doctors, and third-party services), costs of materials, and communication agency costs have to be anticipated. A gant diagram can help to define the success and difficulties that the team can face during the MOOC. To prevent such difficulties, the importance of the use of clear deliverables (synopsis/storyboards) of a clear system of videos postproduction validation, a communication agency services, and transcript society have to be explained and followed by all the members of the team.

For the evaluation part of the MOOC, 1774 participants were registered in total. The platform can provide figures that include the total number of active learners, those who watched the videos, those who tried quizzes, and those who participate in the forums. Those graphs give a percent per week.

A questionnaire was given to the learners at the end of the MOOC to evaluate their experience with the MOOC.

Managing a MOOC with doctors is very particular. Many legal aspects have to be cleared before. Time is also the sinews of war because doctors are most and foremost practitioners in hospital, and they have a limited time to dedicate to the project.

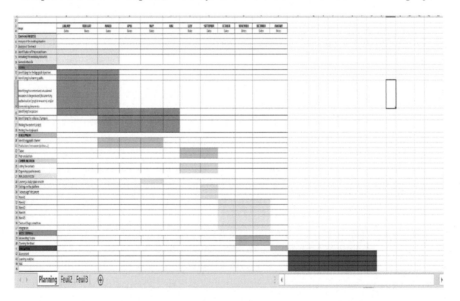

Six months were necessary to design the MOOC, 3 months to shoot and record videos, 2 months to implement the videos in the platform, and 2 months to advertise the MOOC.

The MOOC benefits from two grants from the ICAN (Institute of cardiometabolism and nutrition) and the UNF3S. The ICAN contributes to advertise the MOOC by financing the communication agency with a 17,000 euros grant, and the UNF3S gave 9000 euros helping to finance the cost of a community manager, and part of this money helped to finance the communication agency.

Sorbonnne Université has also a dedicated budget for MOOCs. Eight hundred euros were given to the society in charge of the subtitles.

Evaluation of the MOOC (key figures)

In total, there were 1774 participants in the MOOC and 300 active learners. In this paragraph, we will present the main figures of the MOOC.

Here are some graphs giving data about learner's traces:

– About the contents (general overview):

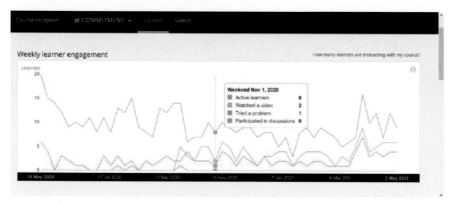

Engagement of the participants per week.

Those measures include active learners, those who watched the videos, those who tried quizzes, and those to participate in the forums.

– About videos (number of views):

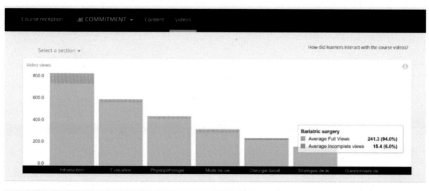

Video views

Order	Section name	Videos	Average Full Views	Average incomplete views	Percent of Completion
1	Introduction	1	750.0	96.0	88.7%
2	Evaluation	4	580.5	27.8	95.4%
3	Pathophysiology of obesity	5	432.6	23.0	95.0%
4	Way of life	7	306.9	31.0	90.8%
5	Bariatric surgery	7	241.3	15.4	94.0%
6	Management strategies	4	158.0	21.8	87.9%
7	Satisfaction questionnaire	-	-	-	-

Show 10 entries   ‹ Previous **1** Next ›

Engagement of the participants per week and per section

Those graphs illustrate per week the percent of videos completely watched and videos incompletely watched.

– About quizzes:

Submission of sections

Order ▲	Section name ⬍	Problems ⬍	Average Correct ⬍	Incorrect Average ⬍	Average number of submissions per issue ⬍	Correct Percent ⬍
1	Introduction	1	686.0	302.0	988.0	69.4%
2	Evaluation	4	617.3	69.5	686.8	89.9%
3	Pathophysiology of obesity	6	395.8	100.5	496.3	79.8%
4	Way of life	5	336.2	19.2	355.4	94.6%
5	Bariatric surgery	7	261.1	22.1	283.3	92.2%
6	Management strategies	2	197.0	15.5	212.5	92.7%
7	Satisfaction questionnaire	-	-	-	-	-

Show 10 ⌄ entries   ‹ Previous **1** Next ›

Engagement of the participants (quizzes)

Those graphs illustrate per week the average percent of participants who complete the quizzes or not.

Evaluation of the MOOC

The questionnaire was divided into three parts: sociodemographic questions (gender, education, and country of origin), general questions on the MOOC (public, social media, contents…), and questions on the pedagogical aspects of the MOOC (quality of the videos, activities, exercises…).

Sociodemographic questions (gender, education, and country of origin)

With no surprise, 90% of the participants came from France and a few from the United States, Switzerland, and Algeria. Approximately 54% of the participants had higher education (masters and Phd), the rest of them were undergraduate. Approximately 30% of the participants were dieticians, 10% psychologists, 6% students, 1% pharmacists, 1% patients or a keen, 1% physiotherapist, 1% sports coach, etc.

General questions on the comprehension of the MOOC

About 70% of the participants found the pedagogical contents adapted to their initial needs. The other 30% claim the needs on more physiological aspects. Those suggestions could be added to the next MOOC. About 60% of the learners follow the MOOC to increase their professional skills, 15% just by curiosity and for general information, and 25% for research purposes.

About 90% of the participants attended the MOOC from the beginning to the attend. We had no droppers. About 50% of the participants heard about the MOOC through social media, 10% through the online newspapers, and 20% by word-of-mouth referrals.

Questions on the pedagogical aspects of the MOOC

Ninety-four percent of the participants measured the pedagogical contents between good and excellent. Approximately 53% of them found the activities (exercise and QCM) of a good quality.

Evaluez la qualité de la pédagogie (vidéos, supports, explications) dans ce cours ?

52 réponses

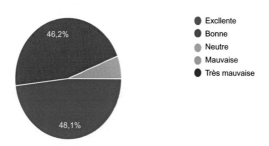

- Excllente
- Bonne
- Neutre
- Mauvaise
- Très mauvaise

46,2%

48,1%

Regarding the use of the forum, 80% found the forum helpful because it helped by giving details and explanations on obesity as a disease. To the question: what are, for you, the professional or personal benefits of this MOOC?

Many of them expressed the fact that they learn more about obesity, they have a positive experience and a good general overview, they improve their skills and knowledge about obesity, and they can talk about obesity in their different aspects.

Overall, participants are satisfied and learn a lot about obesity. Many of them claimed the fact that some subjects have to be detailed more, especially on surgery or more details on the psychological aspects. There is a real motivation from the team to take into account those suggestions for the next sessions.

Conclusion

The MOOC was launch on January 24, 2018. The team faced some difficulties, especially running out of time. It asks the questions of doing an MOOC when you hold a medical activity that give you no time to spend on a long project such as an MOOC. The doctors were not prepared to face this amount of work. They realize that creating an MOOC is not just transforming your power points on some videos. It needs to rethink on their pedagogy and their way of teaching. Feedbacks from discussion forums tell us that learners are mainly professionals and the questions asked are medically orientated.

Case study number 2

Educational projects at CAPSULE follow a defined process. The teacher must be able to complete a form describing his project, its relevance, and its impact on teaching.

The project is then examined by the management office and assigned to a pedagogical engineer according to his availability and appetence for the subject whenever possible.

The projects are evaluated according to their degree of hybridization, the current share of face-to-face teaching, and the extent to which we wish to pass on to distance learning. This is the beginning of the work between the educational engineer and the teacher who will guide him in the completion of his project.

In this green chemistry project, we enter into the "nonexistent presential" scenario of the competice model presented earlier.

Description of the project

It is about a complementary open and remote course of 1 ECTS (out of pedagogical contract) with an hourly volume of 10-h conferences/seminars. It is aimed at L2 and L3 chemistry undergraduate students and aims to raise students' awareness of sustainable chemistry.

Students will be self-initiated into green chemistry and sustainable chemistry with the help of online videos illustrating the 12 founding principles of green chemistry. This self-training will be supplemented by conferences involving specialists in the field (researchers and industrialists). A minimum of two lectures will be necessary to reach the first level of the green label. Students will only travel for the final evaluation. Moodle allows the follow-up of the students.

Method: This is a new course designed to interest chemistry audience in their university curriculum. The project leaders propose an offer to students of certified training (Green Labels 1 and 2) as a complement to their course. The "added value" of this project is to enable the learning public to acquire a collective awareness that can be valued as citizens or future chemists.

Illustration: This course available in initial training was set up by three chemistry teachers. A lecture theater presenting the course was organized by the teachers to open the course and explain to the students the objectives of the training. The course consists of two modules present on Moodle: A module of introduction and initiation to green chemistry, its principles and indicators proposing a biblioverte (documentary resources), a video in motion design, and random calculation questions. *Students are divided into groups*. They have access to the same resources (videos) randomly distributed over 10 weeks and will have to answer comprehension questions about the videos they visit. The students can, thus, consult all the resources at their own pace. A discussion forum has been opened for each group and listed through the different resources to offer students the opportunity to discuss the proposed topics. A virtual class is available in the middle of the semester to review with the students their misunderstanding and answer any outstanding questions. The scripting of the course was conducted with the pedagogical engineer of the project. The birth of the project lasted 2 years to create a motion design video resource with the help of a graphic designer. This pilot video allowed to define the identity of the project, the logo, to allow the teachers to become familiar with the storyboard, the writing of a script, and the graphisms to be integrated. This introduction video is the first of a series of 12 remaining videos to be produced. For the moment, the materials are in French, as well as the communications, but eventually they will be offered in English. At the end of the semester, two mandatory face-to-face conferences followed by a questionnaire on the content of the conferences will validate the course.

Evaluation conditions: To validate the course, students will have to answer the Introduction to Green Chemistry quiz (part 1), which will count for 15% of the overall mark. This quiz will consist of 10 questions with 3 attempts in limited time. The grade will be the average of the 3 attempts, answering the quiz on the different themes (part 2), which will count for 35% of the global grade. The quiz will consist of 20 questions.

In summative evaluation and following the two lectures, students will answer 20 questions in a limited time in the classroom (1 h). Room numbers will be communicated to you later. This last test will count for 50% of your overall grade.

Conditions for success: (1) pedagogical: teachers must be reactive and provide a response within a contextualized time frame. (2) organizational: a strong partnership between teacher and pedagogical engineer is necessary (3) technical: Moodle is used and maintained and the contents (sequences and pedagogical supports) are continuously improved.

According to the grid of the University of Laval, here is an analysis of the blended learning course of green chemistry:

Items	Description	Green chemistry
Course overview and presentation	– The course and how it works are presented to the students: – Explain the general functioning of the course, including the purpose, general objectives, and the place of the course in the program(s), the methods of supervision, the course schedule (by week or by content), the learning activities to be conducted, and the methods of evaluating learning. – Add explanations and illustrations as needed. Examples: On the home page, a short explanatory text or a guided tour of the different sections of the lesson plan and the course site. In the "Course description" section, well-filled headings.	**Course title**: Green Label **Teaching mode**: blended learning **Credit**: 1 ECTS **Positioning**: Complementary course integrated into the chemistry degree and requiring basic knowledge in chemistry **Course content:** The course is spread over 12 weeks with a progressive opening of the modules and will cover the following notions: – Introduction to green chemistry, its principles and indicators (open from February 28 to March 28). Only resources will be available until May 26. – Introduction to some major issues in the field of sustainable development (open from March 28 to April 28). Please note that once the "Introduction to Green Chemistry" block will be open, it will be available until the end of the semester. **Aim of the course**: At a time when plastic is found both in the Marianas trench and at the summit of Mount Everest and when toxic products, suspended particles, global warming, and the depletion of oil reserves weigh on the future of the planet, choosing this complementary training offer will enable you to acquire a collective awareness that you can value as a citizen or future chemist.
General and specific objectives	General objective (linked to specific objectives) At the end of the course, the student will be able to:	General objective (linked to specific objectives) At the end of the course, the student will be able to: – Understand the 12 principles of green chemistry and apply them in the laboratory and in daily life. Specific objectives: By the end of this module, the student will be able to: Know how to calculate the E factor How to calculate the EA factor – List the 12 Principles of Green Chemistry and understand their objectives.

Assessment of learning	The proposed evaluations are consistent with the specific objectives The proposed assessment methods are consistent with teaching/learning methods. The evaluation methods are multiple and are distributed throughout the session The assessment instructions and correction criteria are clear. The procedure for dealing with delays in the handover of work is made explicit. Periodically, the teacher provides students with formative assessment activities with feedback and comments. The different accepted file formats are specified in the instruction.	To validate the EU, you will have to: – Answer the Introduction to Green Chemistry quiz (part 1), which will count for 15% of the overall mark. This quiz will consist of 10 questions. You will be allowed three attempts per question in a time limit of 20 min. Please note—your score will be the average of your three attempts. – Answer the quiz on the different themes (part 2), which will count for 35% of the overall mark. This quiz will consist of 20 questions. You will be allowed three attempts per question in a time limit of 20 min. Attention—your score will be the average of your three attempts. – Attend the two proposed lectures (attendance required) on the morning of May 26. Following the two lectures, you will be invited to answer a 20-question quiz in a limited time in the room (1 h). The room numbers will be communicated to you later. This last test will count for 50% of your overall grade.
Teaching and learning activities	The choice of educational activities and resources is consistent with the general and specific objectives. The course is divided into modules in a coherent, logical, and balanced way. Each module includes a presentation, objectives, pedagogical activities, and learning resources. The course includes a variety of activities Activities support active learning: they are motivating, engaging, and being interactive. For each module, the list of actions to be conducted by the students is clearly stated. The learning activities are of increasing complexity in relation to the course objectives. The weekly student workload is equal to the number of credits in the course. The teacher invites the students to suggest areas for improvement in the teaching activities.	Students will learn about green and sustainable chemistry through a variety of resources in English and French: – online videos, – excerpts from books – video and text quizzes (about 30 questions), – a Biblioverte (documentary resources...), – a forum: This will allow you to ask questions to your teachers and peers. We strongly recommend you to be active in this forum. This self-training will be complemented by two conferences given by specialists in the field (researchers, industrialists). These conferences will take place in May 2020.

Continued

Items	Description	Green chemistry
Student Interactions	Activities that encourage interaction between students are proposed Forums or other spaces for exchange and discussion are proposed Collaborative work tools are offered to students During the exchange and discussion activities, student participation is clearly stated The rules for the creation of teams are clearly stated It is suggested that students introduce themselves to their colleagues and share their expectations of the course. Various activities are offered to students to develop their sense of belonging to the group	– A forum: This will allow you to ask questions to your teachers and peers. We strongly recommend that you be active in this forum.
Terms and conditions of supervision	The teacher gets in touch with the students in the first week and introduces him/herself. The role of the teacher, if any, of other stakeholders involved in the course is explained to the students. Communication methods and the time frame for answering students' questions are clearly indicated in the course outline. The deadline for correcting assessments is indicated. The teacher participates in and guides the discussions. Answers to students' questions are accurate and complete. Teacher helps students stay motivated and engaged. The teacher accompanies active or struggling students by making interventions to engage and involve them. Through explanations and comments, the teacher helps the students to progress. At the end of the course, the teacher gives the students an opportunity to reflect on their learning.	An introductory lecture theater course will take place to introduce the course and its issues. The teacher will regularly answer students' questions on the forum and will organize a virtual class in midterm to answer six specific questions from students requiring a particular understanding.

Category	Criteria	Notes
Use of Technology	Institutional tools are preferred The use of external sites are referenced and available No broken links Various tools are used	The course will be available on Moodle University's LMS. External resources (videos) have been selected.
Educational Resources	Teacher explains what the proposed resources will be used for and how to use them. Resources are up to date. The course makes use of a variety of resources including open-access resources. A clear distinction is made between compulsory and optional teaching resources. The bibliography of the course is up to date. The educational resources are based on various media (text, sound, image, video, etc.). The deposited text resources are up to date, clear, and structured. The presentation of the resources is neat and legible. Works used in the production of multimedia material are clearly referenced. The teacher invites students to report any problems related to resources and access. The course provides resources to help the students become more successful.	All videos are mandatory to watch and completion tracking is enabled.

Conclusion

The profession of an educational engineer is constantly evolving and adapting to the specificities of institutions and establishments. It is a field job that requires an open-mindedness and a permanent monitoring of the tools and associated practices. Above all, he or she acts as an interface between the teaching teams for the design of training systems in line with university policy. The educational engineer does not work alone. They call upon different resources: developers, video artists, and graphic designers.

They are often members of a community of educational engineers and of a community of practice and must be trained and kept informed.

The strong impact of training systems such as distance learning and hybridization make the pedagogical engineer an essential task force in his expertise in supporting teachers in the success of pedagogical projects.

The unprecedented situation of COVID-19 that we are presently experiencing will definitively transform our teaching practices, which are still cautious in certain disciplines. Vigilance will be focused on the ability to create true hybrid courses and not a simple transposition from the classroom to the distance.

The educational engineer will not stop training and developing skills in innovative practices: virtual reality, immersive games, etc.

References

[1] Batier C, Bettler E, Coltice N, Douzet C, Docq F, Lebrun M, Letor C, Liétar A, Burton R, Mancuso G, Rennoboog E, Gueudet G, Lameul G, Morin C, Nagels M, Albero B, Eneau J, Borruat S, Charlier B, Rossier A, Deschryver N, Peraya D, Peltier C, Ronchi A, Villiot-Leclercq E. HY-SUP, http://prac-hysup.univ-lyon1.fr; 2012.

[2] Bocquet F, Bouthry A, Duveau-Patureau V, Haeuw F, Lesage M-C, Perrey P, Pouts-Lajus S, Roy-Picardi D, Schaff J-L. Competice, https://eduscol.education.fr/bd/competice/superieur/competice/libre/index.php#; 2010.

[3] Contamines J. Guide de conception d'un scénario d'apprentissage. INF9013 Les TIC et l'apprentissage en milieu de travail, Université TÉLUQ; n.d. [En ligne].

[4] Aubry-Abel C, Belzile B, Bernier C, Boyer R, Bouchard-Vincent T, Casista É, Cloutier M, Cormier É, Croteau-Bouffard M-H, De Koninck Z, Desroches L, Dorais S, Dorval G, Dumais M, Dubé A-V, Duenas AP, Ebacher M-F, Gérin-Lajoie S, Gervais F, Gilbert D, Larrivée S, Laroche A, Marquis F, Martel É, Martin N, Picard M, Potvin C, Proulx J, Proulx I, Rivest S, Samson L, Sénéchal J-F, Turpin D. Guide des bonnes pratiques de l'enseignement en ligne Partage d'idées, conseils et suggestions, https://www.enseigner.ulaval.ca/guide-web/guide-des-bonnes-pratiques-de-l-enseignement-en-ligne#categorie-164; 2021.

[5] Leroux JL. Un cadre de référence pratique pour soutenir l'évaluation des apprentissages à distance. University of Sherbrooke; 2019 [en ligne; page viewed on April 2020] https://www.fadio.net/wp-content/uploads/2019/02/Cadre-de-r%c3%a9f%c3%a9rence-pratique-pour-soutenir-l%c3%a9valuation-des-apprentissages-%c3%a0-distance_-Leroux_JL_FADIO_LerouxJ_L_et-collaborateurs_2019-02-20VF_lien_Fiche-pratiquepptx.pdf.

[6] Mayer RE. Applying the science of learning: evidence-based principles of the design of multimedia instruction. Am Psychol 2008;63(8):760–9.

Further reading

Chabatura J. Using bloom's taxonomy to write effective learning objectives. Tips for UofA Faculty University of Arkansas; 2020 [en ligne; page viewed on April 2020] https://tips.uark.edu/using-blooms-taxonomy/.

Lauer R. Ingénierie pédagogique et modèle ADDIE. Sydologie Le magazine de l'innovation pédagogique; September 2017 [en ligne; page viewed on April 2020] http://sydologie.com/2017/09/ingenierie-pedagogique-modele-addie/.

Chalifour C. Le zoom pédagogique: les courants pédagogiques. Pari (Université de Bordeaux); 2019 [en ligne; page viewed on April 2020] https://www.parilab3e.fr/le-zoom-pedagogique-les-courants-pedagogiques/.

Marcelle Parr M, editor. Pour apprivoiser la distance, Guide de formation et de soutien aux acteurs de la formation à distance. Refad; March 2019 [en ligne; page viewed on May 2020] http://www.refad.ca/wp-content/uploads/2019/05/Pour_apprivoiser_la_distance__-_Guide_de_formation_et_de_soutien_aux_acteurs_de_la_FAD.pdf.

Caroline L. L'hybridation dans l'enseignement universitaire pour repenser l'articulation entre cours magistraux et travaux dirigés. Ripes 2020;36(1) [en ligne; page viewed on April 2020] https://journals.openedition.org/ripes/1067?lang=en#tocto2n2.

Utilizing the power of blended learning through varied presentation styles of lightboard videos

Christoph Dominik Zimmermann[a], Alvita Ardisara[b], Claire Meiling McColl[c], Thierry Koscielniak[d], Etienne Blanc[e,f], Xavier Coumoul[e,f], and Fun Man Fung[g,h]

[a]*Department of Materials Science and Engineering, National University of Singapore, Singapore, Singapore*
[b]*Department of Food Science and Technology, National University of Singapore, Singapore, Singapore*
[c]*Department of Sociology, National University of Singapore, Singapore, Singapore*
[d]*Conservatoire National des Arts et Métiers (Le Cnam), DN1 – Direction Nationale des Usages du Numérique, Paris, France*
[e]*University of Paris, UFR Biomedical Sciences, Paris, France*
[f]*INSERM UMR-S 1124, T3S, Paris, France*
[g]*Department of Chemistry, National University of Singapore, Singapore, Singapore*
[h]*Institute for Applied Learning Sciences and Educational Technology (ALSET), NUS, Singapore, Singapore*

Introduction

The blended learning approach

"Blended learning" has been popularized in many higher education circles as a method of integrating instructional modalities, methods of learning, and offline-to-online approaches to teaching and learning [1]. This approach focuses on the use of computer-based technologies such as video production, as explored in this study [1]. The blended learning approach has radically transformed the traditional, face-to-face learning systems and the way educators interact and engage with students. Ultimately, this approach satisfies the growing expectations and demands for higher-caliber learning experiences and results by both educators and students in today's advanced higher education environments.

Technology-Enabled Blended Learning Experiences for Chemistry Education and Outreach
https://doi.org/10.1016/B978-0-12-822879-1.00003-2

Computer-based technologies: Video production with the lightboard

The computer-based technologies used in blended learning to stimulate learning both in and out of the classroom include videos in the form of self-paced, web-based video courses and live e-learning videos [2]. The popularization of video production used in education (especially the flipped classroom style of videos) has led to the invention of the lightboard to solve the challenges of a traditional whiteboard setup for educational videos.

Pioneered by Northwestern University Professor, Michael Peshkin, the lightboard is a glass writing board used for producing instructional video lectures where the lecturer stands front-facing toward the camera and annotates the lightboard with handwritten notes and diagrams. Most importantly, the front-facing feature of the lightboard setup allows lecturers to maintain eye contact with the camera and write their notes in front of them, unobscured by his/her body. According to Guo, this visual representation provides good circumstances for creating engaging and interactive educational content. The lightboard technology combines the effectiveness of Khan Academy-style video production and the direct engagement and interaction of the lecturer with the students engaging with the video content.

A pedagogical advantage of the lightboard is that it boosts learning and engagement through the delivery of short, easily retainable bits of information. This is seen when the lecturer has to end the video when the lightboard is fully annotated with charts, equations, and notes. Students can then process the information presented on the board, hence limiting the amount of information a student receives in a single sitting. As such, various learning institutes have started to popularize the use of lightboard videos, be it in flipped classroom or in their normal classes as supplementary learning materials [3, 4].

A lightboard setup (Fig. 2.1) consists of a mounted black backdrop, an LED-lit glass board, a monitoring device, and video recording device. A teacher then stands behind the lightboard and delivers his or her lecture uses various colors of fluorescent markers to aid in writing, drawing, sketching, and annotating the lecture material.

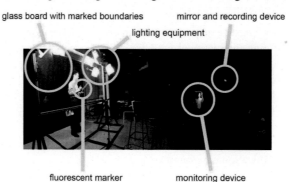

FIG. 2.1

Setup of a lightboard at the National University of Singapore, Centre for Instructional Technology (NUS CIT).

FIG. 2.2

Lightboard video showing an instructor writing down a concept on the glass board in
(A), two instructors explaining the lightboard in (B), and lightboard video with a diagram
overlaid and annotated in (C).

Currently, a lightboard video comprises of one (Fig. 2.2A) or two instructors
presenting a topic behind the glass board (Fig. 2.2B). Sometimes, graphics are overlaid and annotated on these videos (Fig. 2.2C) in real time using a video mixer [5].
However, there is a gap in the literature exploring lightboard videos and varied presentation styles besides the one mentioned earlier. There has also been a gap in evaluating the effectiveness of different video styles in viewer engagement [6–8].

For this reason, this preliminary study experiments and evaluates three unique
video styles using the lightboard as a tool for educational videos. The ultimate aim
of the report is to evaluate which style of lightboard videos is best suited for student
engagement and learning experience the overall blended learning pedagogy.

Materials and methods

A lightboard was previously setup by the National University of Singapore, Centre
for Instructional Technology (NUS CIT) according to step-by-step instructions provided by Peshkin in lightboard.info [9]. Fluorescent dry erase markers (Expo Neon)
were used to annotate on the glass board.

Three different video styles were produced: *interview style*, *multipresenter*, and
multimedia-enriched. These three styles are illustrated in Fig. 2.3 and explained in
further detail in Table 2.1. The style of each video consists of up to four lecturers
demonstrating a concept in the field of chemistry or biochemistry to a student or a

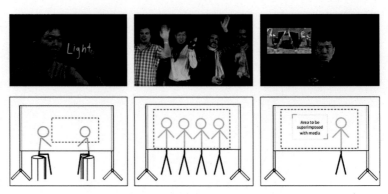

FIG. 2.3

Top, from left to right: Interview style, multipresenter, and multimedia-enriched lightboard videos. Bottom, from left to right: Schematic diagram of the recording setups for each video style. Presenters are positioned behind the glass board and annotate normally on the board, that is, not writing backward. Dashed lines on the diagram refer to the camera frame representing the area around the lightboard that is captured on video.

guest (Fig. 2.4). During the recording, the camera focuses solely on the presenter/interviewee's visage, whereas the interviewer's face remains out of the camera.

Before the production of each lightboard video, lecturers craft scripts pertaining to biochemistry topics and would rehearse these script in front of the camera to familiarize themselves with the content, delivery, and verbal cues before the final recording [10,11]. Depending on the scope, depth, and plan for the content delivered, some lecturers choose to annotate the glass board before recording with keywords and the concepts covered during the video. Additional preparations include cleaning the glass board to prevent ink smudges from being visible on camera. During recording, lecturers use different colors of fluorescent markers to write on the glass board to highlight different text hierarchies or to signify different ideas and concepts. Overall, preparation and recording of a lightboard video takes 1–2 h of teaching time. Table 2.1 shows an overview of how each video style (interview, multipresenter, and multimedia-enriched) was recorded.

Results and discussion

In this section, we explore the results of the three different video styles, *interview style*, *multipresenter*, and *multimedia-enriched*, by examining the pros and cons of each video style. Following this, we provide a possible solution or recommendation to the disadvantage(s) of each method.

Interviewer style
Pros
The interviewer style creates a more interactive, two-way conversation between a presenter and an interviewer as opposed to the conventional, single-presenter

Table 2.1 An overview of how to record each video style (interview, multipresenter, and multimedia-enriched).

	Interview style	Multipresenter	Multimedia-enriched
Number of presenters	1–2	2–4	1–3
Method	Presenter(s) and interviewer(s) are each seated on a stool placed at close proximity to the glass board, ensuring that the speakers can write on the glass board without having to lean forward	Presenters rehearse their movements and blocked their positions behind the glass board	Presenters mark an area on the glass board where they will paste any multimedia files, for example, photographs and videos. During postproduction, the multimedia files will be overlaid on the recorded video at the marked area. If further annotations are made within the boxed area, during postproduction, the following is conducted: After superimposing the media, the original lightboard clip is duplicated and pasted on top of the media. If Final Cut Pro or Adobe Premiere is used, the blending mode of the duplicated video is changed to screen or lighten. Otherwise, the opacity of the video is reduced
Camera framing	Camera zoomed in to frame the presenter in a talking head format	Camera zoomed to frame all presenters in the video	Camera zoomed to ensure everything that needs to be recorded is in frame
	The interviewer may or may not be captured on camera	All presenters are captured on camera	All presenters are captured on camera

lightboard videos. With an interviewer supporting the dialog, the presenter receives verbal cues and prompts in the event where the presenter forgets their script or has lost their train of thought. This style of video production increases the confidence of the presenter because the human aid provided by the interviewer reduces the overall awkwardness normally felt when recording a conventional straight-to-camera lightboard lecture video.

FIG. 2.4

Photograph of interview style setup. During the recording, the camera is zoomed in to show only the presenter/interviewee's visage. The interviewer's face remains out of the camera.

Cons

However, the effectiveness and quality of work from the interviewer style of video production depends on both the interviewer and the presenter being prepared or confident of the video content. In addition, this style may not be the most efficient for a presenter because the question-and-answer interview format requires more time and repeated discussion between the presenter and interviewer rather than a smooth, one-way delivery by a single presenter in a lightboard video. The length of the video production process is increased if the presenter is easily swayed by digression.

Another drawback cited by a presenter when testing this style of video production was the difficulty in multitasking between writing notes on the lightboard and attending to the interviewer's questions. He also noted that sitting down restricts the presenter's range of movement, hence reducing the amount of text one can write on the lightboard. Understanding this limitation, one improvement that we propose is for both the interviewer and presenter to stand up, allowing a full range of movement and space to write on the lightboard. This slight variation in the interviewer style results bears close resemblance to the multipresenter style of video production.

Multipresenter style

Generally, there are only slight distinctions between interview style and multipresenter style of video production. Nonetheless, the multipresenter style, as the name suggests, requires two or more presenters to engage in interactive conversations with each other and the camera. The dynamics of content delivery is further increased because the presenters are not bound to taking turns to speak on camera; instead, they are able to engage semispontaneous conversations to engage their audiences further.

Pros

Similar to the advantages of interview-style videos, having multiple presenters allows for mutual support of delivering accurate and a comprehensive scope of lecture content because both presenters support each other in recalling scripts and video content through verbal cues and collaboration of ideas. This advantage assumes that all presenters have a degree of self-control and expertise over their content and are not easily swayed away by digressions.

Cons

Because the multipresenter video style requires two or more presenters present and visible in the video frame, it results in minimal space available for both presenters to move and annotate on the lightboard while staying within the frame. In addition, if the number of presenters increases, this style of video production is severely compromised because visibility of annotations on the lightboard becomes increasingly challenging to comprehend, the amount of space allocated to each presenter for writing and moving becomes limited, and, ultimately, the sheer number of moving and talking parts in the frame distracts viewers from the annotations on the lightboard and the essence of the lecture content. This is further aggravated when one or more presenters wear light-colored clothing, against the text in light, fluorescent colors, which makes annotations on the lightboard less legible (Fig. 2.5).

FIG. 2.5

When the presenter *(right)* wears light-colored clothing, annotations on the lightboard are less legible.

Measures to counter these limitations would be to limit the number of presenters to less than four people and to encourage all presenters to wear dark-colored clothing. This would provide sufficient space for each presenter to move more freely in the frame and to avoid distractions and poor visibility of the annotations on the lightboard.

Multimedia-enriched style

The last video presentation style is a multimedia-enriched lightboard video.

Pros

This style is useful when a presenter requires complex diagrams for explaining concepts such as protein structures. As exemplified by Peshkin, this style of video production requires one to superimpose lightboard videos with media postproduction. This process is relatively simple and does not require additional video equipment such as a video switcher or mixing of video sources to allow for real-time annotation of diagrams.

Cons

The main challenge of using the multimedia-enriched style of presentation is the accuracy of the placement of annotations because the presenter has to estimate where they need to annotate on the lightboard. This requires preproduction strategic planning on the part of the presenter because he/she has to be conscious of the placement of any superimposed diagrams.

Another drawback of using this style of video production is that the presenter is unable to make annotations and notes using fluorescent markers, most commonly used in lightboard videos. This is because technical requirements of blending the superimposed images on the lightboard require the media used to be in dark colors. Hence the annotations would not work on light-colored backgrounds. One suggestion to reduce this limitation would be to add both the media and additional annotations during postproduction editing. However, these solutions fundamentally defeat the purpose of Lightboard videos because its aim is to minimize postproduction editing in the production of lecture videos. With these limitations, it appears that the multimedia-enriched video style would be the most impractical for use to produce a lightboard video. Still, it would prove useful in cases where no further annotations are needed on the superimposed media.

Conclusion

This study explores how the different lightboard production styles—interviewer, multipresenter, and multimedia-enriched, generate various modes of instructional teaching to enhance the blended learning approach to self-paced, web-based video courses and live e-learning videos. Despite each style producing minor limitations, these exploratory experiments show the multitude of possibilities for using video technology through the lightboard to enhance higher learning.

Acknowledgments

The authors thank the USPC-NUS grant for innovation in higher education in supporting Technology-Enabled Blended Learning Experience #TEBLE. Grant number 2018-02-EDU/USPC-NUS "VIPER: Virtual reality and Innovative Pedagogy in EnRiched environment." We wish to thank the NUS CIT for the continuous support and Dr. Mariana Losada for her contribution in strengthening exchange between Singapore and France. In 2017, this lightboard project "Curating Lightboard Videos for Better Student Engagement in a Large Flipped Classroom Course" was shortlisted for Digital Contents Award from QS-STAR Reimagine Education.

References

[1] Bonk CJ, Graham CR, Cross J, Moore MG. The handbook of blended learning: global perspectives. Local Designs, San Francisco: Pfeiffer; 2012.
[2] Singh H. Building effective blended learning programs. Issue Educ Technol 2004;43(6):51–4.
[3] Ye W. Lightboard and Chinese language instruction. J Technol Chin Lang Teach 2016;7:97–112.
[4] Smith T, Knight C, Penumetcha M. Lightboard, camera, nutrition! J Am Acad Nutr Diet 2017;117(9):A70.
[5] Peshkin M. 5. Switcher. In: Lightboard; 2014. [Online]. Available: https://lightboard.info/home/software.html. [Accessed 21 February 2021].
[6] Tune JD, Sturek M, Basile DP. Flipped classroom model improves graduate student performance in cardiovascular, respiratory, and renal physiology. Adv Physiol Educ 2013;37(4):316–20.
[7] Strayer JF. How learning in an inverted classroom influences cooperation, innovation and task orientation. Learn Environ Res 2012;15:171–93.
[8] Guo PJ, Kim J, Rubin R. How video production affects student engagement: an empirical study of MOOC videos. In: Proceedings of the first ACM conference on learning @ scale conference; 2014.
[9] Narayanan S. Lightboard. National University of Singapore; 2017. [Online]. Available: https://wiki.nus.edu.sg/pages/viewpage.action?pageId=135954958. [Accessed 21 February 2021].
[10] Seery MK. ConfChem conference on flipped classroom: student engagement with flipped chemistry lectures. J Chem Educ 2015;92(9):1566–7.
[11] Fung FM. Adopting lightboard for a chemistry flipped classroom to improve technology-enhanced videos for better learner engagement. J Chem Educ 2017;94(7):956–9.

Further reading

Rossi RD. ConfChem conference on flipped classroom: improving student engagement in organic chemistry using the inverted classroom model. J Chem Educ 2015;92(9):1577–9.
Luker C, Muzyka J, Belford R. Introduction to the Spring 2014 ConfChem on the flipped classroom. J Chem Educ 2015;92(9):1564–5.
Pienta NJ. A "flipped classroom" reality check. J Chem Educ 2016;93(1):1–2.

Trogden BG. ConfChem conference on flipped classroom: reclaiming face time—how an organic chemistry flipped classroom provided access to increased guided engagement. J Chem Educ 2015;92(9):1570–1.

Shattuck JC. A parallel controlled study of the effectiveness of a partially flipped organic chemistry course on student performance, perceptions, and course completion. J Chem Educ 2016;93:1984–92.

Gloudeman MW, Shah-Manek B, Wong TH, Vo C, Ip EJ. Use of condensed videos in a flipped classroom for pharmaceutical calculations: student perceptions and academic performance. Curr Pharm Teach Learn 2018;10:206–10.

McLaughlin JE, Griffin LM, Esserman DA, Davidson CA, Glatt DM, Roth MT, Gharkholonarehe N, Mumper RJ. Pharmacy student engagement, performance, and perception in a flipped satellite classroom. Am J Pharm Educ 2013;77:196.

Guy R, Marquis G. Flipped classroom: a comparison of student performance using instructional videos and podcasts versus the lecture-based model of instruction. Inf Sci Inf Technol 2016;13:1–13.

Squire K. Video game-based learning: an emerging paradigm for instruction. Perform Improv Q 2013;21(2):7–36.

Squire K. Video games and education: designing learning systems for an interactive age. Educ Technol 2008;48:17–26.

Stieff M, Werner S, Meador D. Online prelaboratory videos improve student performance in the general chemistry laboratory. J Chem Educ 2018;95(8):1260–6.

Jordan JT, Box MC, Eguren KE, Parker TA, Saraldi-Gallardo VM, Wolfe MI, Gallardo-Williams MT. Effectiveness of student-generated video as a teaching tool for an instrumental technique in the organic chemistry laboratory. J Chem Educ 2016;93(1):141–5.

Boevé AJ, Meijer RR, Bosker RJ, Vugteveen J, Hoekstra R, Albers CJ. Implementing the flipped classroom: an exploration of study behaviour and student performance. High Educ 2017;74:1015–32.

Ryan MD, Reid SA. Impact of the flipped classroom on student performance and retention: a parallel controlled study in general chemistry. J Chem Educ 2016;93(1):13–23.

Curriculum design, implementation, and evaluation, outreach

Using mobile phone applications to teach and learn organic chemistry

J.L. Kiappes[a,b]

[a]*Department of Chemistry, University College London, London, United Kingdom*
[b]*Corpus Christi College, University of Oxford, Oxford, United Kingdom*

Introduction

University students often encounter organic chemistry early in their studies, either as an independent course or as part of general or GOB courses. Organic chemistry requires the student to learn new vocabulary and factual information, cultivate conceptual understanding but also to combine and apply these in higher-order ways to interpret how and why reactions happen. Students often find the balancing of multiple considerations required for reaction mechanisms particularly challenging; organic courses are often identified as "gatekeeper" courses, with such high rates of withdrawals and failing grades that, after the course, many students change their major to different STEM subjects or, indeed, away from STEM entirely [1]. Because of this, substantial effort and research have aimed to develop a variety of tools and methodologies to make organic concepts as accessible as possible but also in ways that maintain student engagement and build student confidence.

Since the "opening" of the first major App Stores in 2008, smartphones and other tablets have become increasingly common place. In 2019 3.2 billion people (approximately 42% of the world population) owned a smart phone [2], with much an even higher proportion of smartphone owners among 18- to 34-year-olds [3], the demographic containing the majority of organic chemistry students. Given the availability of and student familiarity with these touchscreen devices, a new canvas in which to develop organic chemistry resources has emerged. Recently, a number of smartphone applications specifically designed to support the teaching of organic chemistry have become available [4,5]. At the same time, other applications developed for more generic purposes (e.g., chatting, electronic whiteboards, and even games) can be employed to enrich student learning in the organic chemistry classroom as well as blended and online learning contexts.

For the purposes of this chapter, applications will be discussed according to the following categories:

1. Visualization
2. Multiple Choice Questions

3. Open-Ended Problem Solving
4. Collaboration

Visualization applications focus on creating videos or interactive models that represent molecular structures and interactions, often in three dimensions. Although these applications might be employed by an instructor to pose a question or students to respond to one, the applications are not centered around set tasks.

Both multiple-choice question and open-ended problem-solving applications are task-oriented, with questions set either by the app developer or course instructor, differing in the style of question. Broadly speaking, both classes of task-oriented applications share several pedagogical underpinnings: providing immediate feedback to students, ways of monitoring progress, and gamification of the learning process.

Finally, collaboration applications are defined here as those that promote social learning, allowing students to work together either synchronously or asynchronously. Most often, these are applications not specifically for organic chemistry, so are adapted by the instructor. These can enable gamified learning similar to the task-oriented applications but also have further scope to be used for discussion and metacognition. This consideration of learning can either be about material presented in lecture or in conjunction with the other classes of apps to help maximize their benefit to students.

The varying aims of the developers allow each type of app to be integrated into organic courses in different roles: dissemination of information, active learning, review, and more. In this chapter, we will consider the capabilities and features of representative applications from each group, discussing methods of implementation in an organic course, limits, and how these limits can be parlayed into learning opportunities.

Visualization applications

Habraken astutely observed that "Chemists cannot talk to each other without the use of drawings" [6], which is even more true in the case of organic chemistry. Students have to become familiar with a variety of two-dimensional representations (wedge-hash, Newman, and Fischer projections; ball-and-stick model; and space-filling model) but also adept at visualizing the three-dimensional interpretations of these images. Given the interplay between structure and reactivity, it is critical to provide students with tools that develop visuospatial reasoning, which is often a shift for students from the more mathematical focus of General Chemistry courses.

Traditionally, molecular model kits serve as the main tool for students to gain hands-on experience with molecular shape and stereochemistry. The ability to independently manipulate perspective and molecular torsion angles often represents a turning point for student understanding over the course of the first semester. Although the model kit can help clarify concepts of chirality and torsion angles, many students still have difficulty relating the three-dimensional model to the ways it is represented on paper or other two-dimensional media. Some projections, like that of Newman, can be observed readily because there is a clear instruction to look down a specific bond, whereas the angle from which to observe a three-dimensional chair conformer to see the standardized two-dimensional projection can be difficult to articulate verbally. Smartphone applications represent a huge opportunity to bridge this gap because the screen shows something in two dimensions, but the touchscreen enables the viewer to manipulate the same in three virtual dimensions—something that can only be imagined with these projections when written on a page or whiteboard.

ModelAR and WebMO are two applications whose foundation is building ball-and-stick electronic models analogously to a physical model kit—selecting individual atoms and then bonding them together. During this building stages, ModelAR places chemistry-based fail-safes such as limiting the number of bonds a particular atom can make and restricting bond angles to optimized values (or as close as possible in constrained situations such as cyclopropanes). By contrast, WebMO allows the molecule to be drawn without these limitations; there is a feature to subsequently "clean up" the structure, which optimizes bond and torsion angles. The "clean up" adds an appropriate number of hydrogen atoms to any hypovalent nuclei but does not correct "Texas carbons" or other hypervalent species, simply adding a corresponding (though not necessarily chemically reasonable) charge to the atom. For example, cleaning up a hexavalent carbon provides a central carbon atom with $+2$ charge and octahedral geometry. Both these systems have benefits. The guidelines of ModelAR help the student to develop a sense of what is possible and appropriate geometry. On the other hand, WebMO requires the student to bring this knowledge with them and actively decide about the reasonableness of structure.

With the structure in either app, the user can rotate, translate, and scale the molecule using the touchscreen so as to view it from a variety of perspectives similar to a physical model. In this way, the representation provided by both applications is more powerful than a static two-dimensional drawing. Furthermore, the applications overcome a limitation of physical models. A molecule made with a physical model kit can be handed from one person to another, but in this handover, the perspective of the model giver is not transferred to the model recipient. With the application, one user can identify a particular perspective of interest and then hand the device to their classmate or student and be certain that they are observing it from the same point of view. With this combination of capabilities, these applications represent a powerful way to help students become more adept at translating between two- and three-dimensional representations.

Research by McCollum et al. [7] supports the benefits that this bridging technology provides. Students were taught about molecular geometry using a physical molecular model kit together with either (1) ball-and-stick models printed on paper or (2) the same structures rendered in a visualization application similar to ModelAR and WebMO. The students were then given three tasks. The first required them to match a given two-dimensional ball-and-stick model to its physical model counterpart; the second to match a given two-dimensional ball-and-stick model to a matching structural formula; and finally, to build a physical model based on a structural formula. In the first two exercises, the ball-and-stick models were provided in the same format as the students were taught in, whereas in the third task, there were no ball-and-stick models available (so no students had access to the application for this tasks). In all three tasks, those who got trained using a touchscreen application were more successful by a statistically significant margin. Although physical molecular model kits will continue to play a key role, this is strong evidence that incorporating these manipulable interfaces into topics such as molecular shape, conformation, and stereochemistry are likely to cultivate the ability of students to conduct the same manipulations mentally.

Beyond this shared feature of providing manipulable ball-and-stick models, ModelAR and WebMO each have additional capabilities. In terms of purpose and functionality, ModelAR serves a very similar role to a physical model kit. Once the molecule of interest is made, the user can rotate about individual single bonds to examine different conformers rather than these being locked into the lowest energy torsion angle as in WebMO. The AR in ModelAR's name is an abbreviation for augmented reality, and users can view their molecular model "in the real world" using their device's camera and a printable anchor symbol available from the developer, Alchemie. This anchor helps provide an alternative method to control perspective beyond the touchscreen. The user can walk around the anchor to similarly walk around the molecule or alternatively rotate the anchor itself so as to rotate the perspective. This method of "looking around" the molecule might be more intuitive for some students, so the choice of methods makes the application more useful to a wider audience. A similar augmented reality model has been applied by other groups to extend the concept to other systems: BiochemAR [8] allows users to interact with potassium channel structures; rather than static molecules, NuPOV (nucleophile's point of view [9]) employs augmented reality to consider the three-dimensional requirements of reactions, specifically the angle at which a nucleophile approaches a carbonyl.

Although WebMO might not be as suitable for considering conformations due to the locked torsional angles, its link to computational chemistry packages makes it appropriate to several other components of the organic chemistry curriculum (Fig. 3.1). After the structure is "cleaned up," the application calculates the point group of the molecule, and the symmetry elements can be superimposed onto the structure individually or all at once, while maintaining the ability to manipulate perspective, to gain a view of the symmetry element from all angles. This visualization of symmetry elements can support instruction of stereochemistry, particularly the types of symmetry common in *meso* compounds (Fig. 3.1a and b). In addition, these symmetry representations are a useful tool during the study of NMR spectroscopy to identify equivalent nuclei when predicting or assigning spectra. In another link to spectroscopy, the WebMO calculates vibrational modes providing both the expected IR absorbance as well as a looped video of the molecular motion.

An acronym within the WebMO name indicates a major feature—the ability to calculate molecular orbitals, visualizing not just the physical arrangement of the molecule but also its electronic structure. A list identifies the energy and occupancy of each molecular orbital, allowing the identification of the frontier molecular orbitals (Fig. 3.1d). These together with electrostatic potential maps (Fig. 3.1f) allow students to link structure with sites of nucleophilicity and electrophilicity.

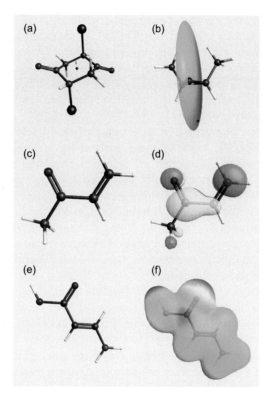

FIG. 3.1

Capabilities of WebMO. Beyond touchscreen models of organic molecules, WebMO can indicate the symmetry elements of the molecule as well as calculate information, including molecular orbitals and electrostatic potential maps. All images in this figure were created with the WebMO app, www.webmo.net. (a) The structure of the meso compound (2R,5S)-2,5-dibromocyclohexane-1,4-dione with its inversion center (i) indicated. (b) The structure of the meso compound (2R,3S)-2,3-dimethyloxirane with its mirror plane (sigma) indicated. (c) The structure of but-3-en-2-one. (d) The LUMO of but-3-en-2-one, with the largest coefficients near the carbonyl carbon and terminal alkene carbon atom, indicating the positions that are most electrophilic. (e) The structure of (E)-3-aminoacrylic acid. (f) The electrostatic potential map of (E)-3-aminoacrylic acid superimposed onto structure, showing the reduced electron density around the heteroatoms donating electron density through resonance compared with the oxygen atom withdrawing electron density through resonance.

In contrast to the previously discussed applications that focus on the three-dimensional structure and properties of molecules, Animator by Alchemie showcases dynamics, allowing even a user without any animation experience to create short videos that demonstrate interactions and reactions. As with the other visualization applications, the user builds molecules by bonding individual atoms together from the virtual molecular model kit, and the arrangements can also include other species such as free ions. With all the relevant species prepared, the user can proceed to produce the video in a style similar to claymation. The species are arranged into position, and the arrangement saved as a frame. The process is repeated for each important "landmark" frame of the video, and Animator automatically connects these frames with smooth movements. For example, to create a video of an S_N2 mechanism, only three frames are required: starting materials, transition state, and products, but the video demonstrates the entire process.

The application can enhance lectures (whether face-to-face or recorded online) with videos specific to the learning goals of the lecturer, as opposed to a more generic library of videos that are provided alongside a textbook. When showing the video to students, it can be presented in its entirety, but the original frames also act as bookmarks, allowing particular features to be emphasized. Continuing the example of an S_N2 reaction mechanism, the first frame exemplifies the archetypal backside approach of the nucleophile, whereas the second spotlights the pentavalent transition state. The same video can be adapted to demonstrate, for example, how a protic solvent solvates the nucleophile, occluding its approach of the electrophile.

Although the process of video production within the app is similar to the step-by-step approach of an electron-pushing mechanism, the video demonstrates the overall flow of the process, giving students a better grasp of the reality of how a reaction takes place compared with static individual steps. Particularly in cases such as an enzyme-catalyzed mechanism, this blending of steps can highlight how hydrogen bonds in an active site that orient a substrate to initiate a reaction can evolve into acid-base catalysis later in the mechanism. This can make clear the insight that these are two ends of a spectrum rather than distinct interactions.

As well as a tool for the instructor to convey information to students, the application is as an alternative to pencil and paper for students to propose mechanisms or explain intermolecular forces. Similar to the analog medium, the student has freedom in which to compose their answer, and the response reflects the student's understanding because the application does not have any "check" or clean up functions for charge and valence (as seen in WebMO and other programs such as ChemDraw). Students enjoy the activity, finding it more creative than writing a mechanism. Especially if the student produces a recorded or written commentary to supplement the video, we have found it to be an effective way to gauge student understanding.

Task-based applications
Gamification

Although feedback and progress are well-established concepts familiar to instructors, gamification is a concept whose popularity has increased in tandem with smartphone application accessibility. Gamification is defined as "the use of design elements characteristic for games in non-game contexts" [10]. Although this term has become *en vogue* only during the past decade, the concept itself has been applied to learning and teaching, and organic chemistry specifically, much earlier with both analog games and earlier types of electronic interfaces [11].

Analog games continue to be an area of innovation for teaching organic content [12], although the majority of organic chemistry applications more closely mirror video games than board or card games. Students, in the mindset of playing a game, find the experience "fun and engaging," and, indeed, engagement is the primary aim of this tool. In addition, the low-stakes game context allows students to receive feedback on their learning and comprehension without feeling graded. Points, scores, and high scores are pillars of gaming that tie into the other aims of the applications, providing information about progress in parallel to the feedback. The format also allows for playful competition to be introduced, with some applications even including formats that allow synchronous competition.

Multiple-choice question applications

The two chemistry-specific multiple-choice question applications presented here were each developed by university-based academics.

Chirality-2 [13], developed by a team at RMIT University, presents a set of six levels that can each be approached independently, in any order, with each connecting to topics typically taught early in an organic chemistry course. The arrangement for the functional group, intermolecular forces, and isomers levels are all similar. The player is presented with a bank of options that are the same for all questions as well as a figure containing chemical structure(s) that need to be labeled. In the functional group game, the player assigns functional group names from the option bank to the groups that are already highlighted on a natural product or drug molecule. The intermolecular forces level highlights interactions which the player then identifies with the most appropriate choice from the bank. Two structures are presented to the player in the isomers game, and the player selects the term that most accurately describes the relationship between them. Structure classification presents a memory-matching style of game but with all cards shown. The player has to match structural formulae to a description, again testing functional groups, along with classification of 1°, 2°, and 3° groups. In the chiral centers game, there is a circle next to each carbon atom of a structure, and the player taps all chiral carbons within the molecule, then predicts the total number of stereoisomers possible. Finally, the naming molecules level provides four possible names for each structure shown, from which the player chooses the correct name. Based on the performance in the game, the player is awarded a medal-stylized benzene ring for each game. As well, the player can check their best score and time for each game as well as their performance (score and time) for the most recent 50 times they have played each level, allowing them to monitor progress over time.

Organic Fanatic [14] truly has the feel of a game, with visuals recalling arcade and computer games of the 1980s and early 1990s. Developed as a collaboration between the Department of Chemistry and the Department of Theater, Film, Television, and Interactive Media at the University of York, the game can be played in either single-player or multiplayer mode. In either mode, the player(s) answer questions, each with four choices, which span many aspects of an introductory organic course: nomenclature (matching a name and structure), predicting products and mechanism (e.g., selecting which intermediate is or is not found as part of a specific mechanism), spectroscopy (e.g., choosing the typical IR absorbance for a functional group), and specific knowledge (e.g., selecting which compound is a component of vinegar). In the single-player mode, the questions are divided by functional group. The player chooses the order in which to answer the functional groups, one question each, with difficulty increasing as the player progresses. If the player chooses a wrong answer, they receive a time penalty but cannot move on to the next question until they have selected the correct answer. At the end of the nine questions, the game informs the player of their total time (including penalties), with the aim to finish as quickly as possible. In multiplayer mode, up to six people play using a single device, passing it around between each question. The interface for each question (including time penalties for wrong answers) is the same, with the game working like Hot Potato—the player does not want to be the one holding the device when time is up. This continues until one player remains as the winner.

As apps specifically designed for students of organic chemistry, both are very well suited for self-assessment and review. The designers of Organic Fanatic specifically had end-of-course review in mind. As such, every game (in either single- or multiplayer mode) includes material from across the entirety of a typical course, limiting its utility for earlier in the year. Conversely, Chirality-2's focuses on foundational topics that allow students to select which topic they will play serves as a useful checkpoint for mastery of key concepts early in the academic year. In addition to the multiplayer mode of Organic Fanatic, healthy competition can be encouraged between students by providing a venue (e.g., an online discussion board) to compare best times and scores in the single player games. Both games draw questions from banks of questions, which is useful for review as a player is less likely to encounter the same question twice, even over a series of several games. However, because the questions are selected at random each time, it is not feasible to have all students answer the same questions in the app at the same time, for example, during a lecture as a synchronized check.

To allow all students to answer the same multiple-choice question at the same time, more generic applications can be employed, with two popular examples being Kahoot! And Quizlet. Kahoot! focuses on the live aspect and provides a wider variety of questions beyond multiple-choice. In particular, a "puzzle" type allows a question that requires players to put several elements into correct order. This can be to either arrange mechanistic steps so that they lead from starting material to product. It is also well-suited to trends such as carbocation stability and nucleophilicity. Finally, some questions can have typed answers—either those that are checked for correctness or truly open-ended allowing for real-time feedback sessions. Because only live sessions are available (the instructor posts a question either sharing online or projecting in the classroom), Kahoot! is ideal for in-class use [15] but not a tool appropriate for independent revision, and its utility for chemistry assessment is unclear [16]. Quizlet has the ability to be used both for in-class, synchronous attempts at the same question as well as independently by students to study and review. Other than multiple-choice questions, it includes flashcard capabilities, with the benefit of a spaced repetition system to help maximize efficiency.

Both Kahoot! and Quizlet allow access to libraries of questions that others have made public, but it becomes a responsibility of the instructor to curate an appropriate bank of questions. This comes with the benefit of tailoring it to the specific course being taught, with a Kahoot! quiz for each specific class meeting or a Quizlet deck for each unit. Furthermore, the flexibility to author questions presents an opportunity for students as well. Asking students to prepare questions based on the material at the end of each unit challenges them to engage critically with the material. Writing multiple-choice questions has them contemplate likely misconceptions and errors. This extends the indirect use of the application to incorporate further aspects of the Bloom taxonomy than required to answer questions.

Open-ended problem-solving applications

As with the multiple-choice question applications, the open-ended ones have a structure such that the application itself sets the goal. However, these applications allow more freedom in the method of response, allowing the students to explore intellectually as they journey toward their answer. Rather than a question-and-answer, this format promotes a more constructivist approach. Paralleling our consideration of the multiple-choice applications, we will look first at those that are designed specifically for organic chemistry.

IsomersAR and Isomers are produced by Alchemie, the same company that produces ModelAR discussed earlier. IsomersAR is similar in interface to the augmented reality mode of ModelAR, with appropriate bond angles shown; however, rather than simply a virtual model kit, IsomersAR has the aim of building all hydrocarbon isomers that contain up to 10 carbon atoms. Whenever a new isomer is produced in the three-dimensional visualizer, it is unlocked in an index that shows the skeletal structure and IUPAC name.

Although IsomersAR uses exploration to "discover" new isomers, Isomers might be better described as a puzzle game. Carbon atoms appear on the screen and cannot be translated. They have to be connected following certain rules (bonds cannot overlap other bonds or carbon atoms) to meet the goal of the individual puzzle. The carbons are arranged such that the puzzle can be solved but not such that bond lengths are uniform or that bond angles match what would be predicted based on chemical principles.

The goals can be defined in several ways. In the most straightforward, the target skeleton itself is provided. Alternatively, the goal might be given as a specific number of primary, secondary, tertiary, and quaternary carbon atoms. In the connections mode, the goal is put in terms of which types of carbons (in terms of degree of substitution) are bonded to each other (e.g., cyclohexane would be the solution to "6 bonds linking 2° to 2°"). Finally, the player can choose to receive the IUPAC name as the goal. Although all modes are enjoyable challenges, the name mode has the most direct application to course content as a way to consolidate nomenclature.

Chairs was the first application specifically designed for organic chemistry education [17]. The aim of the game is to promote understanding of chair flips of cyclohexanes and how these are represented on chair representations in two dimensions. The interface is straightforward: the player is presented with a substituted cyclohexane chair and the unsubstituted, flipped chair. Positions on the unsubstituted chair are highlighted where the player uses the touchscreen interface to add the substituents at the appropriate angle. The student is given feedback during play with the substituent bond turning green when at the correct angle, even before the player commits to the answer. A timer bar counts down, with time added for every correctly placed substituent, allowing the game to go indefinitely if the player is able to keep placing the substituents correctly in a timely enough fashion. The game starts with monosubstituted structures, adding a further substituent every fifth chair. As well as gaining time for each correctly placed substituent, the player gets a point, and the game keeps a record of the high score.

When used in a college course in a flipped-classroom setting, students who used the game performed better on a paper-based quiz on chair flips at the end of the class compared with a control group who did not [17]. Furthermore, students who used the app reported that they found it helpful in visualizing cyclohexanes. Since recommending the application in classes for the past several years, I can anecdotally report that I often hear students comparing high scores (and asking what mine is for comparison!). The application is particularly useful because the task it focuses on is one that can be greatly improved with practice but requires feedback for students to know that they are doing it properly. Whereas an instructor would have limited time to give feedback, the application allows students to have corrections on unlimited examples and attempts. Thus, they make the most of any synchronous or face-to-face time by seeking more specific help from the instructor.

Mechanisms [18] is an application that addresses head-on three of the five most difficult organic chemistry concepts (as reported by organic chemistry educators [19]): reaction mechanisms, acid-base chemistry, and resonance. Not only are these concepts challenging for students, but they are also key foundations for deeper understanding of and further study in chemistry. Mechanisms was developed as a way to allow students to practice reaction mechanisms in an active way that provides formative feedback that is feasible in classes of all sizes, even those so large (> 500 students) where provision of consistent, formative feedback on a large number of examples is logistically challenging.

Incorporating feedback from more than 20 educational institutions in 8 different countries throughout development [18], Mechanisms includes more than 250 mechanistic puzzles that span from resonance and acid-base chemistry to that of a variety of functional groups encountered in a first-year organic chemistry course. The application uses the touchscreen to drag and move electron pairs. The curly arrow formalism is not explicitly introduced or required for the puzzles, but the moves employ the same logic. One major advantage is the open "sandbox" nature of the game that allows students to move any pair of electrons any direction (i.e., students can make "wrong" moves). Scaffolding within the game immediately alerts the student to pursue an alternative route when a wrong move is made, whereas the same mistake might cause them to lose a lot of time when attempting the mechanism on paper.

Scaffolding is a key aspect of the Mechanisms app [20] (Fig. 3.2). At the start of each puzzle, the player is shown a panel that shows either a key intermediate or the curly arrows of a single step from some part of the mechanism to be performed (Fig. 3.2a). From there, the player is then presented with the interactive starting materials; any atom can be tapped to show attached hydrogens and lone pairs (Fig. 3.2b). Chemical reasoning can be used in the same way as solving a mechanism with pencil and paper. If an incorrect move is made (Fig. 3.2c), the involved atoms will take on a spiky appearance and the move will be reversed. In some cases, a warning sign appears (Fig. 3.2d) to provide the player a hint explaining why the move is unlikely (Fig. 3.2e). If one move requires another as part of the same step (e.g., the concerted nucleophilic attack and leaving group departure in an S_N2), the app requires the steps to be carried out in the more likely order (the attack first to avoid formation of the primary carbocation), then darkens the background to indicate a further move is necessary (Fig. 3.2f), with the background returning to normal once an intermediate is reached. If the player is stuck at any point, "Goals" (Fig. 3.2g) are available for each puzzle, which provide information about the product or key intermediates.

Mechanisms gives users the opportunity to select which specific puzzle they would like to attempt, and instructors can apply for an account that allows them to assign particular puzzles to a class or section. In this way, Mechanisms can easily be incorporated as an interactive aspect of a lecture, an entire active learning session of a flipped class or independently as homework or review. The application also can be integrated with Epiphany, an interface that allows the instructor to assign specific puzzles as well as see student attempts and success. This makes it easier for an instructor to distribute subsets of puzzles for in-class activities, homework, or even examinations.

As with the visualization applications, it is worthwhile to discuss the ways in which the touchscreen approach of Mechanisms influences how students approach and solve a mechanistic problem in comparison to the traditional handwritten approach. Finkenstaedt-Quinn and colleagues, using think-aloud interviews, reported similarities and differences in student thinking for mechanistic questions, specifically acid-base [21] and addition to carbon-carbon π-bonds [22]. In both these studies, students generally focused on explicitly given guidance (scaffolding as described earlier in the case of the application and chemical formula of the product in the case of pencil and paper). The differences for both types of reactions hinged on the fact that the Mechanisms application redirected students in cases where they proposed an unfavorable step. In the case of acids and bases,

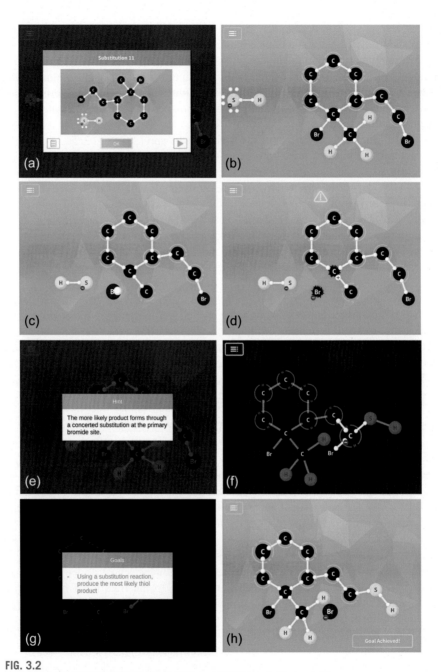

FIG. 3.2

See the legend in opposite page

a prime example is found in the nonequivalent nitrogen atoms of an imidazole ring. Students solving the problem on paper selected one of the two and carried forward with it, whereas students using the application were only allowed by the application to proceed when they selected the more basic nitrogen. Similarly, in the addition of hydrogen halides to double bonds (an asymmetric alkene and an enol ether), students who successfully applied chemical reasoning reached the correct answer in both modalities. Whereas students with fundamental difficulties with bonding followed through an incorrect mechanism on paper, those using the application were shepherded to the correct route through the hints and other scaffolding of Mechanisms. With this in mind, both methods have advantages. The scaffolding of the application can act as "training wheels" for those developing mechanistic skill: it helps to keep the player from "falling over" but does not impede at all those players able to "maintain balance," because they apply chemical principles independently. However, the pencil-and-paper method more accurately represents the situation encountered in proposing a mechanism for an unfamiliar reaction. For these reasons, it is worth incorporating Mechanisms into organic courses while not using it to completely replace the traditional paper-and-pencil modality.

Perhaps most interestingly, the work of Finkenstaedt-Quinn exemplifies that, regardless of modality, it is worthwhile to supplement the tasks themselves with questions about reasoning and encouraging metacognition. Asking students to justify the steps of their mechanisms (whether expressed on a touchscreen or paper) reveals the extent of their chemical understanding and ability to

---

FIG. 3.2

Scaffolding in the Mechanisms application. Other than as an interface to explore electron flow and propose mechanisms, the Mechanisms application provides a framework of information to support and guide users when stuck on a mechanistic puzzle. All images in this figure were created with the Mechanisms app, www.alchem.ie. (a) After selecting a puzzle, the user is presented with a card that reveals relevant lone pairs, a key intermediate or curly arrow formalism for one step of the mechanism. (b) By tapping on an atom, attached lone pairs (as for the thiolate sulfur atom) and hydrogens (as for the methyl group) can be explicitly shown. (c) Students are allowed to move any pair of electrons within the puzzle. Here, the electrons are moved to have the bromide leave behind a carbocation. (d) When the move is unlikely, the atoms involved take on a spiky appearance, and the move is automatically reversed by the game. The yellow "!" attention sign at the top indicates a hint is available regarding the move. (e) The attention sign hint provides formative feedback in response to the specific incorrect move just made by the user. (f) The background darkens to indicate a further pair of electrons must move as part of the same step. (g) In addition to move-specific hints, the user can choose at any time to see the goals for the puzzle that provide more guidance about how to proceed. (h) After achieving all of the goals within the puzzle, the user is commended for their achievement.

apply it. The discussion can act as scaffolding for how to apply chemical principles, further helping students to improve. Recognizing this, the developers of Mechanisms have resources on their website to use to supplement the use of the application itself with further exercises and points for discussion [18]. Whether the application itself is used in class or at home, there are important discussions to be had among students and instructors about the principles that underlie the hints, goals, and guidance of Mechanisms.

In an exciting use of a perhaps unexpected application, Charades!, Koh and Fung have adapted a party game and its associated app into a fun method to review a unit, having selected the specific topic of the chemical laboratory [23]. Charades! is similar to the "Heads Up!" game application made popular on the American daytime talk show, Ellen. One player places the device in front of their forehead with the screen facing outward; the application displays a word to the other players who give clues about the word to the player with the device. The goal of the game is to provide the device holder enough information about the displayed object to identify it. Providing clues to a classmate demands higher-order thinking on the part of the clue givers than would be required for an individual flashcard activity.

The adapted version, ChemCharades, uses a bank of laboratory apparatus and techniques. In addition to the basic mechanics of the game, it is proposed that the device is passed to another player after each term is successfully identified, which promotes equal participation of all students in the device holder and clue giver roles. The device holder is not allowed to pose any questions to the clue givers, whereas the clue givers should use only words, and no gestures, to provide clues. In the original version proposed by Koh and Fung, it is suggested that the instructor observe each round and give feedback after each 120-second round about how more detailed and accurate descriptions might have been given. Students who have received feedback from the instructor can then act as peer facilitators, providing similar feedback to other teams and promoting collaborative learning.

Fishovitz, Crawford, and Kloepper seized the invitation made by Koh and Fung to apply the "Heads Up!" model to other topics, developing games about amino acids, metabolic pathways, and analytical instruments [24]. Although these were not using an application, the parallels to ChemCharades allow some of the developments to be adapted to the application-based game. Rather than focusing on facilitator feedback (from either the instructor or peer), the instructor provides a framework for the types of clues that should be given and in what order. For example, in the case of the amino acids, the proposed scheme was:

1. verbal description of the side chain;
2. class of the side chain (charged, polar, hydrophobic, etc.);
3. pK_a;
4. one-letter abbreviation; and
5. three-letter abbreviation.

The clues allow for a wide range of knowledge and ability. For instance, the verbal description might be as simple as "contains nitrogen" or as specific as "contains a guanidine group." The categories in the framework are increasingly specific, and students are encouraged to give all the clues even if the device holder is able to correctly guess after an earlier detail. Similar to Koh and Fung, the authors noted the utility of the collaborative nature of the game, with several misconceptions being cleared up by peer feedback without instructor intervention.

Similar to Mechanisms, post-activity discussion can enrich the student experience of ChemCharades type games. Because this activity works well for topics with similar or overlapping concepts that might be confused by students, an opportunity for dialog after the game allows students to share insights gained over the course of the game. In particular, discussing which items were the most difficult to describe and to guess can help students identify which areas are mastered or to share advice about challenging concepts in a peer-to-peer context.

Collaboration applications

ChemCharades promotes collaboration as part of the way that players achieve the task set by the application. Now, we consider applications whose purpose is to promote collaboration. As such, none of these applications were specifically designed with chemical education in mind but can be adapted to allow organic chemistry learners and instructors to work together synchronously, asynchronously or both. Rather than each application individually, two specific learning environments will be discussed and the applications employed to create these experiences will be presented in these contexts.

For several years in our first-year organic chemistry course for biochemists at the University of Oxford, we have organized research-centered workshops. Working in small groups (4–6 students), the participants apply concepts and problem-solving techniques from the course to case studies drawn from recent research articles. When pivoting to online teaching in response to the COVID-19 pandemic, we wanted to preserve the key elements of these sessions because they enthused students about the sophisticated questions they could already answer during their first year of university study. Furthermore, the workshops serve as key opportunities for student collaboration. Although instructors (approximately 1 for every 15 students) are present to facilitate, interactions are primarily peer-to-peer, with students resolving misconceptions and discussing the plausibility of proposals as a team. Indeed, these elements formed the three pillars to maintain in an online environment:

- synchronous collaboration between students;
- research-centered questions; and
- tutor support in an unobtrusive way that allows students to take the intellectual lead.

Based on these aims, we wanted to use a conferencing software with breakout room capability, so that we could begin and end the workshop with a joint session but provide for smaller groups for the actual problem-solving portions so that all students could actively contribute. At the time, we elected to use Zoom, but Microsoft Teams has since had the feature added. Moving between the individual breakout rooms was not as natural as circulating between groups at individual tables in a large room. However, students could call the main host to the room using the "Ask for Help," and this information had to be relayed to cohost instructors by the host. As the chat within the call only allows discussions between people in the same room, we found it useful to have an additional chat between instructors external to the call to help coordinate motion between the breakout rooms.

Although Zoom and Teams both provide electronic whiteboards as an internal feature, we elected to use the Miro whiteboard. Because it functions in web browsers, as well as applications on touchscreen devices, Miro allowed participation by students with many different levels of device accessibility. Similarly, the ease of drawing and uploading photographs meant that students could contribute either by uploading pictures of work done on paper or draw electronically if a stylus was available. The whiteboard (with interactive functionality) remained available to students after the session, allowing asynchronous discussion to continue between the group members and tutor. Finally, all the instructors could have all the whiteboards open at the same time, giving a sense of which groups might need additional assistance or scaffolding. This helped to avoid unnecessarily entering all rooms frequently, which was often observed to disrupt the flow of student discussion. Initiatives at other institutions [25], including some implemented in an asynchronous context [26], have employed similar methodologies and likewise realized great success.

Students had previously completed other workshops in person and rated the online workshop similarly useful. Discussion began more slowly during the online workshop, likely because the breakout rooms were assigned randomly, whereas students normally select their team when the workshop takes place in person. However, working with new people was pedagogically valuable because students were exposed to novel perspectives, rather than defaulting to normal roles within a regular study group, and is a change that we think could be worth incorporating even to in-person workshops.

In contrast to the synchronous workshop, Flipgrid allows a different type of collaboration. Again, as part of the pivot to online teaching during the COVID-19 pandemic, many chemistry educators expressed concern about evaluating student understanding of mechanism, rather than rote reproduction. The instructor posts a question to which students produce a video response, which can either be immediately available to the entire class or only to the instructor (and shared more widely later if desired). The medium is an asynchronous option specifically for education, complementing more widely used social media platforms [26], all

Table 3.1 Summary of applications.

Application	Specific to chemistry?	Brief general description	Interfaces available
Animator	Yes	Create animated short videos showing interactions on an atomistic or particulate scale	Smartphone and tablet
Chairs	Yes	A game for drawing in cyclohexane substituents after a chair flip	Smartphone and tablet
Charades!	No	A game where other players describe a term to the device holder to help them guess the term	Smartphone and tablet
Chirality-2	Yes	A game of multiple-choice and matching questions about foundational organic definitions and topics	Smartphone and tablet
Flipgrid	No	A video discussion forum. Instructor posts a question that students answer in video format.	Smartphone, tablet, and web browser
Isomers	Yes	A puzzle game of connecting carbon atoms to build a molecule meeting certain specifications (name, number of 3° carbons, etc.).	Smartphone, and tablet
IsomersAR	Yes	A game in augmented reality, building 3D models of alkanes containing up to 10 carbons.	Smartphone, and tablet
Kahoot!	Yes	An individually developed "pub quiz" style game to be played synchronously. Includes multiple choice, ordering, and type in the answer questions.	Smartphone, tablet, and web browser
Mechanisms	Yes	A game moving electron pairs to solve puzzles, showing resonance and how reactions proceed.	Smartphone, tablet, and web browser
Miro	No	An online white board that multiple users can contributed to in parallel (either by drawing directly on the board or attaching a file). It also has chat capability.	Smartphone, tablet, web browser, and desktop
ModelAR	Yes	A 3D molecular model kit with an additional augmented reality mode.	Smartphone and tablet
Organic Fanatic	Yes	A game with multiple choice questions about the entire subject of organic chemistry, either in single-player mode or "hot potato" style multiplayer.	Smartphone and tablet
Quizlet	No	An individually developed "pub style" quiz with multiple choice questions to be played synchronously. Revise independently with flashcards on a spaced repetition system or other matching games using the flashcards as a basis.	Smartphone, tablet, and web browser
Microsoft Teams	No	An application for voice- and video-conferencing. Includes breakout room capability for small group work.	Smartphone, tablet, web browser, and desktop
WebMO	Yes	A 3D molecular model kit with an ability to analyze point group, display symmetry elements, and calculate molecular orbitals and other electronic aspects.	Smartphone and tablet
Zoom	No	An application for voice- and video-conferencing. Includes breakout room capability for small group work.	Smartphone, tablet, web browser, and desktop

Throughout the chapter, examples of uses of each application have been presented. Some applications such as Animator, Kahoot!, Quizlet, Charades!, and the collaboration applications can be used for a variety of, if not all, topics in an organic chemistry course. However, others are better suited to specific topics. In Table 3.2, units typically covered in an introductory organic course are listed with related apps and suggested uses for how to incorporate them into the classroom.

which are useful for the metacognition discussions proposed in earlier sections for the Mechanisms and ChemCharades activities.

The application can also play a role in assessment, either on its own or as a supplement to a written examination question where students orally explain their answer in a Flipgrid video. For example, elaborating on a mechanism allows the student to show deeper understanding of principles such as nucleophilicity, electrophilicity, and leaving group ability. Furthermore, it provides an alternative method of evaluation, increasing accessibility to a broader spectrum of students.

Conclusions

All the applications discussed in this chapter are summarized in Table 3.1. Beyond the information included in the main text, a list of the compatible devices for each application is included. As discussed in the introduction, smartphones have permeated society to a large degree, which has driven the "application rush." However, as alluded to in the discussion of Miro in the Collaboration section, it remains important to keep in mind that not all students will have access to the same pieces of technology. When teaching occurs in person, this can be addressed with a number of devices available to be loaned to students during the class session. On the other hand, when the assignment is to be completed remotely, asynchronously, or, more generally, independently, we found it a great advantage if the application can also be accessed on a computer (either as an application or through web browser).

Developers, both academics and dedicated companies such as Alchemie, continue to actively pursue innovations in applications designed specifically for the teaching of chemistry. Applications specifically designed for chemistry and organic chemistry arise from collaboration between those with chemical and mobile application expertise, whereas other applications such as Kahoot!, Quizlet, and Charades enable instructors (and students) to create new learning opportunities without specific knowledge of application development. Excitingly, Alchemie is currently developing a new learning platform that combines aspects of their individual chemistry applications, abilities of the synchronous collaboration applications as well as new functionalities. This will allow experiences such as the workshop discussed in this chapter to take place in an environment specifically designed for expressing chemistry.

Although this chapter focused on applications to address theoretical aspects of organic chemistry, the practical laboratory represents a fertile area for application development. Indeed, with the current limitations of social distancing and increased accessibility of augmented and virtual realities, this is definitely a space worth watching.

Table 3.2 Supplementing the organic curriculum with applications.

Topic	Application	Suggested use(s)
Nomenclature and functional groups	IsomersAR	Since this game shows the name as new structures are built by the user, it can be a great introduction to nomenclature. Structures are limited to alkanes, so students can build molecules and note the names provided. In small groups, they can engineer rules based on the patterns they observe to "discover" how nomenclature works for alkanes.
	Isomers	In the name mode, students can test their understanding of nomenclature (either from the IsomersAR activity or otherwise) by interpreting the names into how to link atoms together.
	Chirality-2	Three of the six levels in the game are related to this topic and serve as a good source of independent revision to assign to students. After the levels identifying functional groups and evaluating their degree of substitution, the information can be combined with nomenclature rules to approach the "Naming Molecules" that expands from the alkanes of Isomers and IsomersAR to include molecules with functional groups.
Conformation	ModelAR	Students can build models of alkanes and use rotate about bonds to put the model into the conformation provided as a Newman projection. Conversely, they can build a model and then translate it into Newman and wedge-hash projections.
	Chairs	Use as a flipped classroom session after introducing the concept of cyclohexane conformations. Students can have a tournament starting the game simultaneously and seeing who can last the longest to achieve the highest score.
Stereochemistry and isomerism	WebMO	After building models of a given compound, students can use the point group identifier and symmetry visualizers to identify symmetry elements within the molecule. This can be used to experience the relationship between symmetry elements and chirality.
	ModelAR	Students can build models to investigate the stereochemical relationships between molecules, using the app to translate 2D representations into 3D ones. The app can be used to look in particular at chiral molecules that display axial chirality rather than having a chiral center.
	Chirality-2	Two of the levels will help students to independently check their understanding of isomerism; one focusing on types of isomerism, by applying the concept to identify the relationship between a pair of structures and the other level builds on the activities with WebMO and ModelAR, identifying chirality, but challenges the student to do so on a 2D structural formula.
Acids and bases	Mechanisms	The acid and base puzzles start with autodissociation of water, so can be incorporated into introduction of the topic (either in lecture or with online videos). The more advanced puzzles include molecules with multiple acidic or basic groups to select as the most likely to react; students can use these as a focal point for small-group discussions.

Continued

Table 3.2 Supplementing the organic curriculum with applications—cont'd

Topic	Application	Suggested use(s)
Reaction mechanisms	WebMO	The application can be used in activities to emphasize the link between frontier molecular orbitals and reactivity. When given mechanisms to work through in class (either individually or in groups), students can be asked to use the app to calculate the HOMO and LUMO of specific starting materials and intermediates and link the appropriate orbital to the arrows they propose.
	Mechanisms	The app is useful in a number of roles in the organic classroom. Within a lecture or video, it provides an alternative, dynamic way for the instructor to show electron flow in a mechanism to complement curly arrow formalism. Alternatively, a puzzle can be used midinstruction as a quick check of understanding. Rather than in a lecture-style context, the puzzles can be used as the centerpiece of a problem-solving session, together with follow-up and analysis questions for the students to discuss their reasoning about the mechanisms in small groups. Because specific problems can be assigned, it can also be incorporated as an aspect of formal assessment (either homework or an examination).
Spectroscopy	WebMO	The calculations provided by the app can link to the measurements made in infrared and NMR spectroscopy. Students can be given spectra for structure determination projects and then build a model of their proposed solution in the application. The app-predicted IR frequencies can be compared with the experimental data, which also allows the student to observe dynamically the motion that corresponds to the IR absorbance. For NMR, the symmetry elements in the app can be used to confirm the equivalent environments of nuclei.

Although some applications (e.g., Kahoot! or Flipgrid) can be adapted to many topics, others are particularly well-suited to particular topics. It is this latter group that are included in this table, organized by chemical topic.

References

[1] Weston TJ, Seymour E, Koch AK, Drake BM. Weed-out classes and their consequences. Cham: Springer International Publishing; 2019. p. 197–243. https://doi.org/10.1007/978-3-030-25304-2_7.

[2] Smartphone users worldwide 2016–2021, https://www.statista.com/statistics/330695/number-of-smartphone-users-worldwide/; 2020. [Accessed 25 August 2020].

[3] Smartphone ownership is growing rapidly around the world, but not always equally, https://www.pewresearch.org/global/2019/02/05/smartphone-ownership-is-growing-rapidly-around-the-world-but-not-always-equally/; 2019. [Accessed 13 August 2020].

[4] Naik GH. Role of iOS and android mobile apps in teaching and learning chemistry. Teaching and the Internet: the application of web apps, networking, and online tech for chemistry education. vol. 1270. American Chemical Society; 2017. p. 19–35. https://doi.org/10.1021/bk-2017-1270.ch002.

[5] Winter J. Playing with chemistry. Nat Rev Chem 2018;2:1. https://doi.org/10.1038/s41570-018-0006-x.

[6] Habraken C. Integrating into chemistry teaching today's student's visuospatial talents and skills, and the teaching of today's chemistry's graphical language. J Sci Educ Technol 2004;13:89–94. https://doi.org/10.1023/B:JOST.0000019641.93420.6f.

[7] McCollum B, Regier L, Leong J, Simpson S, Sterner S. The effects of using touch-screen devices on students' molecular visualization and representational competence skills. J Chem Educ 2014;91:1810–7. https://doi.org/10.1021/ed400674v.

[8] Sung R-J, Wilson AT, Lo SM, Crowl LM, Nardi J, St. Clair K, et al. BiochemAR: an augmented reality educational tool for teaching macromolecular structure and function. J Chem Educ 2020;97:147–53. https://doi.org/10.1021/acs.jchemed.8b00691.

[9] Aw JK, Boellaard KC, Tan TK, Yap J, Loh YP, Colasson B, et al. Interacting with three-dimensional molecular structures using an augmented reality mobile app. J Chem Educ 2020;97:3877–81. https://doi.org/10.1021/acs.jchemed.0c00387.

[10] Robson K, Plangger K, Kietzmann J, McCarthy I, Pitt L. Is it all a game? Understanding the principles of gamification. Bus Horiz 2015. https://doi.org/10.1016/j.bushor.2015.03.00610.1016/j.bushor.2015.03.006.

[11] Campbell S., Muzyka J. Chemistry game shows. J Chem Educ 2002;79. https://doi.org/10.1021/ed079p458.

[12] Triboni E, Weber G. MOL: developing a European-style board game to teach organic chemistry. J Chem Educ 2018;95. https://doi.org/10.1021/acs.jchemed.7b00408.

[13] Jones O, Spichkova M, Spencer M. Chirality-2: development of a multilevel mobile gaming app to support the teaching of introductory undergraduate-level organic chemistry. J Chem Educ 2018;95. https://doi.org/10.1021/acs.jchemed.7b00856.

[14] Shoesmith J, Hook J, Parsons A, Hurst G. Organic fanatic: a quiz-based mobile application game to support learning the structure and reactivity of organic compounds. J Chem Educ 2020. https://doi.org/10.1021/acs.jchemed.0c00492.

[15] Ares AM, Bernal J, Nozal MJ, Sánchez FJ, Bernal J. Results of the use of Kahoot! gamification tool in a course of chemistry. In: Proceedings of the fourth international conference on higher education advances; 2018. p. 1215–22. https://doi.org/10.4995/HEAD18.2018.8179.

[16] Murciano-Calles J. Use of Kahoot for assessment in chemistry education: a comparative study. J Chem Educ 2020;97:4209–13. https://doi.org/10.1021/acs.jchemed.0c00348.

[17] Winter J, Wentzel M, Ahluwalia S. Chairs!: a mobile game for organic chemistry students to learn the ring flip of cyclohexane. J Chem Educ 2016;93:1657–9. https://doi.org/10.1021/acs.jchemed.5b00872.

[18] Winter J, Wegwerth S, DeKorver B, Morsch L, DeSutter D, Goldman L, et al. The mechanisms app and platform: a new game-based product for learning curved arrow notation; 2019. p. 99–115. https://doi.org/10.1021/bk-2019-1336.ch007.

[19] Duis J. Organic chemistry educators' perspectives on fundamental concepts and misconceptions: an exploratory study. J Chem Educ 2011;88. https://doi.org/10.1021/ed1007266.

[20] Winter JE, Engalan J, Wegwerth SE, Manchester GJ, Wentzel MT, Evans MJ, et al. The Shrewd Guess: can a software system assist students in hypothesis-driven learning for organic chemistry? Journal of Chemical Education 2020;97(12):4520–6. https://doi.org/10.1021/acs.jchemed.0c00246.

[21] Petterson M, Watts F, Snyder-White E, Archer S, Shultz G, Finkenstaedt-Quinn S. Eliciting student thinking about acid-base reactions via app and paper-pencil based problem solving. Chem Educ Res Pract 2020. https://doi.org/10.1039/C9RP00260J.

[22] Finkenstaedt-Quinn S, Watts F, Petterson M, Archer S, Snyder-White E, Shultz G. Exploring student thinking about addition reactions. J Chem Educ 2020;97. https://doi.org/10.1021/acs.jchemed.0c00141.

[23] Koh S, Fung FM. Applying a quiz-show style game to facilitate effective chemistry lexical communication. J Chem Educ 2018;95. https://doi.org/10.1021/acs.jchemed.7b00857.

[24] Fishovitz J, Crawford G, Kloepper K. Guided heads-up: a collaborative game that promotes metacognition and synthesis of material while emphasizing higher-order thinking. J Chem Educ 2020;97. https://doi.org/10.1021/acs.jchemed.9b00904.

[25] Tan HR, Chng WH, Chonardo C, Ng MTT, Fung FM. How chemists achieve active learning online during the COVID-19 pandemic: using the community of inquiry (coi) framework to support remote teaching. J Chem Educ 2020;97:2512–8. https://doi.org/10.1021/acs.jchemed.0c00541.

[26] Geyer AM. Facebook: an avenue to reflective discussions through case studies. In: Teaching and the Internet: the application of web apps, networking, and online tech for chemistry education, vol. 1270. American Chemical Society; 2017. p. 1–17. https://doi.org/10.1021/bk-2017-1270.ch001.

Interactive and innovative practices to stimulate learning processes in biochemistry

Xavier Coumoul[a,b], **Thierry Koscielniak**[c], **Fun Man Fung**[d], **and Etienne Blanc**[a,b]

[a]*University of Paris, UFR Biomedical Sciences, Paris, France*
[b]*INSERM UMR-S 1124, T3S, Paris, France*
[c]*Conservatoire National des Arts et Métiers (Le Cnam), DN1 – Direction Nationale des Usages du Numérique, Paris, France*
[d]*Department of Chemistry, National University of Singapore, Singapore, Singapore*

Introduction

Teaching has taken many forms in history, from the dialectical style between the teacher and her/his fellows (in Ancient Greece) to lectures in amphitheaters in front of large audiences. The latter form has been traditionally used in universities, but it is hard to imagine using methods on young audiences (e.g., kindergarten). Instead, other techniques are used, such as working in smaller groups (e.g., memorization of letters or numbers). Nowadays, active learning techniques (not necessarily inspired by teaching methods used for children) are introduced by many teachers in universities to ensure the transmission of complex concepts. In this chapter, we will introduce (1) why we thought that new forms of lectures were needed for biochemistry students, (2) the different activities that could be translated to other disciplines, and (3) the perspectives of new developments.

The *motivation* to create new forms of lectures for biochemistry students

Learning processes involve two key partners: the teacher and the student. Although it is obvious that the success of transmission of information between the two depends on the quality of the communication between them, the nature of passing on knowledge has relied on traditional means since the foundation of universities (the middle of the 11th century for Western universities). Most universities still rely on a unidirectional way of transmission: the traditional lecture, whose length varies generally from 50 min to 3 h according to each university [1]. Paradoxically, although this practice appears perfectly normal for both students and teachers (and most of the population), no one outside this context would consider a dialog between two persons as normal if based on a monolog lasting more than 10 min. Interaction seems to be the natural way to learn, and more importantly, to enrich the passing of knowledge.

Despite nearly a millennium of usage, the traditional lecture has come under more and more scrutiny. B.G. Davis, in his book *Tools for Teaching*, states that a period of 10–15 min draws sufficient and efficient attention from the students (but not more) [2]. This statement drew controversy recently because it appears that the quantification of the period had never been properly performed (with methodological flaws and subjectivity in data collection). A recent study showed that scaling attention levels of students over a 50-min course revealed a pattern of increased attention over the first 10–20 min, followed by a steady decline [3]. More importantly, comparing classes, it also revealed that "teaching style" could play a significant role in defining the timing and shape of these two different periods.

At the University of Paris, several biochemistry tenured professors in the early 2000s became aware of the need to optimize teaching practices to promote better assimilation of theoretical and practical knowledge from the students. A key point was the personal assessment by these teachers that the lessons given in lecture halls or amphitheaters contributed to the rapid loss of attention of many students after only 30 min of lesson, confirming the precepts defined previously.

Beyond the search for a better transmission of information by improving the course materials or their practical provision on a learning management system (LNS; e.g., the Moodle platform at University of Paris), the biochemistry and chemistry departments have worked together on how to improve the acquisition of knowledge by the students, during and gradually after the course; they not only worked on memorization through regular learning but also on the consolidation of acquired knowledge. They used different forms of innovative teaching (numerical or nonnumerical). They also tried to improve the confidence of students (e.g., reverse courses adapted to learning science, with an inverted role for the teacher, which benefit from the student's personalized learning, place the students on good tracks for their future professionalization).

Results/conclusion

A necessary dialog between teachers and students during the classroom

Lectures (with potentially 200–300 students) have the major advantage of being able to reach large numbers of students over short periods of time, giving them information without the need for endless duplication. This, however, is also a disadvantage because the enlargement of the classroom constitutes an important dilution factor for the attention of the students [4]. Unfortunately, this is enriched by the inherent length of the lessons (classically 1 h 30 min or 2 h at the University of Paris in biomedical sciences).

In biochemistry, many courses include concepts that require rote memorization: during the first years, for example, students must learn the formulas of amino acids, fatty acids, and sugars, with an additional difficulty related to isomeric forms. These notions are essential for understanding over the following years ("making connections"), the metabolism of these molecules (e.g., glycolysis), which if not presented in a framework integrating physiology and adaptation to cellular needs may also require learning "by heart" (biochemical reactions, name of the enzymes, and effectors).

From this observation, several teachers have gradually implemented a concept of fragmented courses, capable of mobilizing different brain areas of the students (listening, memorization, and reflection) by conducting various activities. A common view of the fragmentation course is presented in Fig. 4.1.

The activities presented in this figure are modular according to the nature of the teaching lesson and discipline. More importantly, they also foster the dialog between teachers and students. Over the variety of activities, two of them can be easily implemented:

Interactive multiple-choice quizzes

With an audience of 200–300 students, introducing alternative activities to interact with the whole amphitheater could represent an attractive way to focus on one major aspect of her/his course. However, it appears clearly that only a few students (the most extraverted ones) will participate. Thus it does not fully represent the present attention of the whole audience [5].

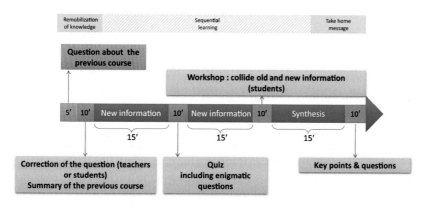

FIG. 4.1

The fragmented course. In this type of course, we use the first 10–15 min to raise the attention of students with activities related to the previous course. The next 15 min are devoted to new information presented in a classical manner. This period is followed by a different activity, such as quizzes (possibly based on enigmatic questions that highlight unexplored parts of the current course). To lever the suspense raised by this activity, the next portion of the course is dedicated to answering some of those questions. Another intercourse period could then be used to ask the students to synthesize, for example, the old and new information. An appropriated synthesis of students' work can then be programmed before answering questions arising from the audience.

During the slideshow, multiple-choice questions can be presented to the students. A short pause (30 s to 1 min) is introduced to let the students think about what the best answer(s) is, and subsequently, the teacher can ask the whole amphitheater the choice of the students simply asking them to raise their hand for each proposal. This technique has the advantage of using only the course material (slide show) and is very easy to set up (and, therefore, still used in practice by several teachers, especially with small enrollments). However, although the participation of the students will be much more important compared with a simple question asked by the teacher, not all the students will participate.

We have, therefore, developed a methodology based on the use of an LNS (e.g., Moodle at University of Paris), which allows the teacher and students to connect during the course and to answer multiple choice questions at different times during the course judiciously chosen by the teacher. To do this, the teachers were trained and supervised by the pedagogical engineers of the university, who also guided us through all stages of development with both technological and logistical activities (see also further). Ninety-six percent of undergraduate students have a smartphone, a tablet, or a personal computer, allowing them to answer these questionnaires live. In this configuration, a proper access of each student to the university network is a key factor for this type of interactive workshop (either through Wi-Fi or a cellular network).

The answers can be displayed on screen after a period of reflection; by experience, this technology achieves much higher response rates than those obtained with freehand responses. In addition, in the case of high percent of errors, we have the opportunity to reset the questionnaire to give students the opportunity to chat with each other to answer the same question again. In general, this additional discussion improves the level of correct answers. Some teachers have also tested software commercialized by private companies (Wooclap, https://www.wooclap.com; Plickers, https://get.plickers.com), which propose nice interfaces to perform such kind of interactive teaching (with many more options).

In conclusion, the number of students engaged in this activity during the course significantly increased (with students who can be physically far from the teacher, for example, at the top of the amphitheater), the answers can be presented only at the end of the activity (preventing students from copying each other), the activity stimulates discussion between students close to each other, and the teacher can interact during the evaluation period to orientate discussions among students.

Interactive workshops

The installation of interactive multiple-choice quizzes led us to reconsider the interaction between teachers and students even further, in the form of a dialog aiming to stimulate learning: during a conversation between two adults whose level of knowledge is different (for example, a teacher and a student), a "questions/responses" dialog stimulates the building of knowledge for the student [6]. Different forms of workshops have, therefore, emerged with an initial question asked by the teacher to the students, a short period of work by the students, followed by a discussion with the teacher:

- The riddles at the end of a course section on which the students must work for 5′
- Puzzles or the use of incomplete images to reorganize, for example, a metabolic pathway (useful in biochemistry)
- The "synthesis" workshop that we will detail here to illustrate this part with an example in the course of biochemistry:

Subject: The role of adenosine triphosphate (ATP), the molecule that symbolizes the energy in the cell) is described on several courses (glycolysis and mitochondrial metabolism). Metabolism is a discipline whose pathways are clearly interconnected to ensure proper regulation of anabolic and catabolic mechanisms into each cell type.

Activity: To show the students that all the courses are connected, the teacher can ask the whole amphitheater to work on a scheme summarizing the origins and roles of ATP. Single students or groups of students (who can have access to the documents from former courses) will then try to build one scheme with their own inputs (memory work or investigating work using the documents).

Teacher work: During these 5 min, the teacher can, of course, provide some hints to any student or groups who asked for help. At the end of the 5-min period, the professor randomly takes four to five schemes, mixes them (to "anonymize" them), and then reads or summarizes on the white/blackboard the different results.

Conclusion: The presentation of data from counterparts eases the peer-review process because the students do have the feeling that potential mistakes can be present. The teacher can decide to let the amphitheater react or ask selected students/groups to provide their feedback (correcting students or groups can be selected at the beginning of the workshop). Collecting such information, the teacher slowly incorporates the good elements for the final correction, explains why incorrect answers from students are…incorrect, etc. The final scheme is finally presented and can be suggested to be part of the essential knowledge for the final examination. This peer-reviewing process helps students with low self-esteem to consider that others can make mistakes and to build confidence. Different supports can be used for this practice: the simple question, the analysis of figures of scientific articles, etc.

One of the major advantages of these techniques is that they rarely require advanced technology for large audiences (more than 200 students); the student only requires a sheet of paper and a pencil, whereas the teacher only needs a blackboard and a chalk. Another advantage is the usage of different brain areas during the course, thus decreasing the monotonic feeling of classical courses. They do not represent a superficial element of the course but again essential knowledge built by the students for their self-examination.

Besides these interactive activities during the classroom, a significant number of teachers felt that interactivity should be developed even further, at home to foster regular learning.

Stimulate regular learning at home

In France, teaching courses in amphitheaters are classically complemented by practical teaching in small classrooms with a restricted audience, allowing for better interaction between students and teachers when working on specific exercises in relation to the main course. However, only a minority of students prepare in advance for such exercises. Regular practice is a key process in the acquisition of knowledge and is at the benefit of the student about timing because it requires less time over whole semesters to perform better for examinations.

Several teaching procedures can be implemented to trigger regular learning among students:

Regular quizzes

Such quizzes can be implemented for students in different areas (enzymology, structure and function of lipids, structure and function of carbohydrates, and bioenergetics). These quizzes can be made available after the course and until the end of the semester, so as to allow students to test their knowledge when they are ready. Each quiz may contain about 20 questions drawn at random from a database, such that the questions will be partly different between two quizzes attempted by the same individual or different individuals. This encourages the multiplication of attempts on the same quiz. The format of the quizzes is variable (multiple choice, true/false, matching, etc.), thus increasing its attractiveness and educational effectiveness. The questions are not timed, and once the student has answered a question, an immediate feedback based on a color code is given so that he/she has immediate feedback on the accuracy of their answer (green: correct answer; yellow: partially correct answer; and red: false answer). The teacher can specify why the answer is correct or not to improve students learning. Each of the 20 questions is marked on 1 point, and after each quiz attempt, a mark on 20 is calculated, complemented with necessary comments to encourage revision. The scores obtained during quiz attempts are not used in the assessment of students; they represent a tool to guide each student in the acquisition of knowledge [7].

Time-limited quizzes

We have also developed such quizzes as part of continuous monitoring in biochemistry. Indeed, some students only choose to learn the course material 2 weeks before the final examination; subsequently, a continuous monitoring was implemented. Participation to this activity improves their final note for the final examination and fosters regular learning by students, while teaching them the advantages of consistent revision of course material over the year.

The principle is as follows: after a course, students have the opportunity to attempt a quiz, available only for a week on the LNS platform, which they can activate when they want. However, unlike the regular quizzes (or the ones used during the course), these are time limited with two attempts [45 s for the first one and 30 s for the second one (if the first fails)]. The time between two attempts is determined by the student, who, in case of failure, can choose to work again on the course (still during the same period of 1 week). The two quizzes are drawn from a random database, which means that two successive attempts do not lead to the same question. These questions are spread over 10 weeks (i.e., 10 quizzes with 1 point obtained for a good answer on the first attempt and 0.5 on the second). The response rates were impressive (90% of participation) and, importantly, last over time with a very low loss of "responding" students (less than 10%).

Other complex modes can be chosen, which alternate learning times (regular quizzes) and evaluation times (time-limited quizzes), following a published three-part course schedule. During the first 3 weeks, we proposed time-limited quiz learning support. After the lecture, the quiz is open for a week. When the student has learned her/his course, he can connect and answer questions drawn at random from a database. In the event of a wrong answer, a message prompts him to rework her/his course and suggests that he come back to assess the quality of her/his learning during the week. In the fourth week, the student will be offered a 4-question evaluation quiz (with no second chance) on the 3-week training program. To perform all these activities, a database of 70 questions for training and 120 questions for evaluation was created.

These time-limited quizzes can represent 100% of the whole assessment or can be partly complemented with examinations. They represent an opportunity to reduce the period of assessment for the teacher, freeing time for other activities. The development of these learning support quizzes has also proved to be very valuable for students enrolled in dual-degree courses (sciences for health mixed with law or economics-management or mathematics or psychology). Indeed, if these students have a well-filled media library of dematerialized course materials, it is not easy for them to organize their learning independently. Training or evaluation quizzes are, therefore, very precious tools. In addition to being effective educational tools, they constitute time markers, allowing them to optimize the regularity of the work for these students with busy schedules (suggesting also that they need continuous examinations to continuously learn and acquire good learning practices). However, they also present the disadvantage to increase basic learning (pure knowledge) versus critical thinking (a personal eye on the acquired knowledge).

Semidigital teaching course

This type of course implemented during the second year at the university aims to facilitate active learning by the students through two distinct patterns:

In the first case, the course builds on the basics acquired by the student during the first year of teaching at the university: a series of quizzes is offered on the Moodle platform, which allow the student to assess her/his level of knowledge acquired during the first year. In practice, the student answers a first series of quizzes (Quiz 1). If she/he reaches a fair level (> 10/20), he can access a second series of quizzes (Quiz 2), higher in difficulty but still based on the supposed achievements of L1. The process continues with Quizzes 3 and 4. All quizzes are made accessible before the start of the course. Thus the first part of the course (20 min) is dedicated to the analysis of the results of the quizzes and continues with a reflection on a scientific problem proposed in advance, in the form of a summary article or a specific scientific question or by a video, before a deepening of the scientific questions is posed within the framework of a lecture.

In the second case, the course brings new information for the student: it then begins with a problem (e.g., deposited in advance on the LNS platform) or a video and a discussion with the teacher. The student then self-assesses her/his understanding of the course through quizzes that he will have to complete in a limited time on the Moodle.

Other practices

- **Maintenance of acquired knowledge**: An extended period between the end of key teaching courses and the final examination (more than 1 month) represents a problem for most students in terms of maintenance of acquired knowledge [8]. To counteract this problem, students were offered the possibility of reviewing and practicing certain course points during the second part of a semester, far beyond the end of the lectures. Weekly, at the end of every class, a video from Khan Academy (https://fr.khanacademy.org) containing a lesson point is proposed. This video is in English with the possibility of subtitling (also in English). The viewing is followed by a question. Subsequently, the student accesses a detailed correction of the question. Thus in the interval between the end of the courses and the examination, the student can see six videos (one per week) containing key points of the course (in English) and progressively work on six different examination subjects according to a schedule orchestrated by the teacher in charge. Of the 80 students enrolled in the teaching unit, 68 participated in the activities (participating can bring up to one bonus point on the examination score). The feedback from students is very positive, in particular through the expression of a feeling of more regular supervision and/ or continuous monitoring that is better distributed over the entire semester. The videos in English, as a complement to the French course, were also very popular and made it possible to kill two birds with one stone: reviewing concepts of the course and improved listening and understanding of scientific English.
- **Interactive-based forums**: The goal of this activity is to enrich the discussion in small practical classrooms complementary to regular courses. Between both activities, a LNS-base forum was opened to allow the students to ask questions about the course, which can be answered by teachers 1 week later. In that particular case, the teachers are able to better anticipate the common difficulties faced by students during small practical classrooms, allowing for greater efficiency in the learning process.
- **Cloud-based peer-to-peer software platforms (such as Zoom or Discord)**: These platforms can be used for distance education and evaluation.

  Such practices gave great results about efficiency to transfer knowledge, keeping the audience physically and mentally attentive during the various classes, and maintenance and acquisition of a large variety of data by the students. However, although learning represents a key element of the university roots, the development of self-criticism and of the own personality of the student is also important for her/his social insertion. Thus we decided to go further in the implementation of interactive activities through the development of personal lectures or workshops by the students.

The development of personal teaching activities by and for the students

The philosophies underlying these practices was (1) to encourage the learning of necessarily more in-depth data with the creation of their own supports (posters or courses), (2) to allow them to better understand the teaching practice and the process of knowledge transmission, (3) to own self-confidence through oral expression in front of various public (other students, teachers, etc.), and (4) to develop their own initiative actions. Over the variety of activities, two of them will be developed here:

(a) **The "fragmented" inverted course**: in the "cell signaling" module in biochemistry, students have created a 2-h course on the AhR signaling pathway (which can be physically separated in two courses of 1 h each). Briefly, the AhR (for Aryl hydrocarbon receptor) is a protein able to detect a large variety of environmental pollutants and to trigger in the nucleus of multiple cells, a transcriptional response allowing the cells to defend themselves against the triggering molecules through the activation of enzymatic process that inactivates them. Beyond the elements that can be developed in biochemistry (protein structure, enzymology, and metabolism) and cell biology (signaling pathway), this course is a good opportunity to link the students to other disciplines such as toxicology and environmental sciences. Because this signaling pathway is activated through a multitude of steps, 10 groups of 5 students ("pentomes") are initially designed and assigned one specific subject corresponding to a step (e.g., the ligands of the AhR; the translocation in the nucleus, etc.). Each "pentome" then prepares a slideshow of five slides and passes in a logical order in front of all the other students, respecting strict formats: five slides max., one student/slide, length of the overall presentation between 5 and 10 min. The students discover each step during this fragmented inverted class, which reconstitutes the overall signaling pathway. The role of the teacher is mainly to modulate the presentations and to give a synthetic conclusion. The final examination includes elements of the presentations by the students and varies each year according to their specificity. The "pentomes" may also receive an extra bonus for the final examination according to the quality of their presentations, assessed through peer reviews. This practice of the "fragmented inverted course" portion of the final examination is now extended, notably within the framework of international teachers/students exchanges, with the National University of Singapore (NUS).

(b) **The peer-review poster creation and presentation**: The biomedical faculty of the University of Paris favored the implementation of transdisciplinary teaching units, including biology, biochemistry, physics, chemistry, and cognitive sciences, offering original courses such as "the DNA: from the double helix to clinical practices." Similarly, the assessment of this course was originally conceived, including a classical final examination (accounting for one-third of the final grade) and the conception of a poster (including a presentation accounting for two-thirds of the final grade, Fig. 4.2).

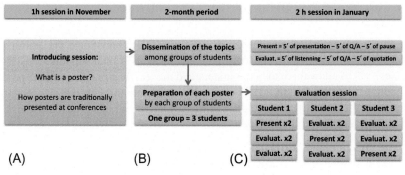

FIG. 4.2

Poster-based evaluation of biochemistry students: (A) the introducing session: a special lecture of 1 h is organized to give them substantive and formal advices on the design and codes that characterized poster sessions displayed in conferences (including the design of a poster) because this is, in general, their first contact with the scientific poster as an element of dissemination and valorization. (B) The preparation of the poster: the students choose their team, but they do not choose the topic of their poster, which is assigned to them randomly (with additional documents such as scientific articles). (C) The evaluation session: several weeks later, the poster session is organized: it lasts approximately 1.5 h for approximately 40 students; during that session, 15-min sessions are fragmented according to the following pattern: 5′ of oral presentation, 5′ of questions and answers, and 5′ of pause and evaluation. Each student presents her/his poster twice (and at least once in front of a teacher) and evaluates the work of his/her classmates four times. The evaluation grid is an app developed in collaboration with computer scientists (CRI for "centre de recherche interdisciplinaire"). Students and teachers use smartphones, tablets, and computers for this task. Interestingly, the poster grade is calculated using the following repartition: 50% of the teacher's grade, 25% of the peer's grade, and 25% by a grade assessing the student's ability to grade similarly to the teachers. In addition to the scientific knowledge acquired during the lectures, the students develop not only numerous transversal skills: teamwork, respect of deadlines, written and oral scientific communication but also the professional competence of an evaluator.

Conclusive remarks

Several tools can be used to stimulate the interactivity with students even if this implicates a large audience, for example, in amphitheaters. Table 4.1 summarized the different tools mentioned in this text together with their advantages and limits.

Perspectives

The philosophy that has been developed within our scientific team regarding innovative and interactive activities to improve teaching, supported by the faculty management and by the university, emerged following observations and discussions with

Table 4.1 The different tools, their advantages, and limits.

Tools	Advantages	Limits
Fragmented course	Keep the attention of students through multiple activities	Requires additional courses to introduce new knowledge (slower pace teaching)
Interactive multiple-choice quizzes ("raise your hand" mode) in front of the students	Better interactivity with the whole audience (+)	Some students might still not participate. Answers influenced by the first students who raise their hands
Interactive multiple-choice quizzes ("LNS" mode) in front of the students	Better interactivity with the whole audience (++); possibility to reset the quizzes and to assess the students through multiple rounds; possibility to use systems developed by private companies (Wooclap); can also be used at home to stimulate regular learning [see also time-limited quizzes ("LNS" mode)]	–
Interactive workshops	Develop critical thinking with the engaged students through questions that go beyond the courses (integration of multiple courses or collateral questioning)	The whole audience might not participate
Time-limited quizzes ("LNS" mode)	Stimulate regular learning with bonus for the final examination; efficient to evaluate the knowledge of the students during lockdown periods	Stressful; do not develop critical thinking; may be perceived as unfair (the questions are randomly selected and their difficulties might vary from one student to another)
Semidigital teaching course or maintenance of acquired knowledge	Build new knowledge with the basics acquired during the first years of teaching	Requires additional courses to introduce new knowledge (slower pace teaching)
Interactive-based forums	Enrich the discussion in small practical classrooms complementary to regular courses	The whole audience might not participate
Cloud-based peer-to-peer software platforms	Distance education and evaluation during lockdown periods	Low engagement by the students. Poor development of critical thinking
Fragmented inverted course	High development of critical thinking, personalized experience, and engagement toward other students	Requires a high level of preparation by the teachers and the students
Peer-review poster	High development of critical thinking, personalized experience, and engagement toward other students	Requires a high level of preparation by the teachers and the students

the students and a rich cooperation with educational engineers; this dialog between engineers and teachers is a key element that allows a perpetual evolution of pedagogic techniques. Inputs by the students have been stimulated by such approaches, including the development, for example, of a YouTube channel that will regularly publish videos on research or teaching projects, as well as portraits of professors or alumni. This philosophy beyond the playful and exciting aspect also leads to the creation of a teaching unit on educational innovation (which aims to have innovative projects developed by undergraduate students supervised by a tutor), the writing of educational articles, and participation in national and international conferences (EDUCAUSE meeting). This leads also to the development of partnerships with other universities, including:

- the NUS to develop other approaches such as the LightBoard (to develop short digital seeds of a format original) or virtual reality activities.
- Unisciel ("UNIversité des SCIences En Ligne," one of the seven French thematic digital universities) as part of a project to create digital seeds (small digital presentations of 5 min), which will be used for a range of practices ranging from the simple revision of courses (on smartphones, for example, in transport) to the creation of more complex video supports (digital inverted courses) by students.

To finish, we believe that these multiple approaches (teaching practices, opening to different disciplines, and involvement of multiple actors, including engineers, students, and partner universities), will help to constantly improve our practices and to discover new ones.

Take-home messages
Target audience
All undergraduate students of the faculty of basic and biomedical sciences and medical students (PACES) are targeted by these teaching practices.
Goals

- Improve knowledge acquisition during the course
- Improve the acquisition of knowledge after the course
- Reinforce the prerequisites (previous years) and acquired (current year) by digital supervised revisions over short periods to also promote regularity in learning
- Involve students in inverting learning

Learning and training settings

- Interactive amphitheater work
- Presentation and oral correction of posters
- Stimulation of regular work at home (and possibly in transport)
- Reverse class to stimulate, beyond learning, students' critical thinking

Involvement of the teaching team, students, and governance

- Development through governance of Wi-Fi equipment in an amphitheater
- Development through the governance of an educational engineering service (PIP, University of Paris) and training by educational engineers on the use of the Moodle platform
- Interactive involvement of students, peer-to-peer evaluation of posters, and reverse courses by students
- Regular organization of educational commissions (every 2 weeks in the presence of student representatives), allowing both adjustments to existing activities and "brainstorming" to bring out ideas

Partnership

- The development of digital seeds is currently being done in partnership with UNISCIEL and the Paris Diderot University
- Development of virtual reality tools is underway with the University of Singapore (NUS)
- The "posters" system involves Antoine Taly and Hugo Lopez from CRI and the Frontières du Vivant license for all the "software and evaluation" development

References

[1] Miller CJ, McNear J, Metz MJ. A comparison of traditional and engaging lecture methods in a large, professional-level course. Adv Physiol Educ 2013;37:347–55.

[2] Davis BG. Tools for teaching. Jossey-Bass; 2009.

[3] Stuart J, Rutherford RJ. Medical student concentration during lectures. Lancet 1978;312:514–6.

[4] Law KMY, Geng S, Li T. Student enrollment, motivation and learning performance in a blended learning environment: the mediating effects of teaching, and cognitive presence. Comput Educ 2019;136.

[5] Gunter GA, Kenny RF. Leveraging multitasking opportunities to increase motivation and engagement in online classrooms: an action research case study. Int J Online Pedagog Course Des 2014;4:17–30.

[6] Hatfield T, Rickley SR. The seven principles in action: improving undergraduate education; 1995.

[7] Guskey TR. How classroom assessments improve learning. (Using data to improve student achievement). Educ Leadersh 1994;60.

[8] Zimmerman BJ, Dibenedetto MK. Mastery learning and assessment: implications for students and teachers in an era of high-stakes testing. Psychol Sch 2008;45(3).

The design of blended learning experiences for clean data to allow proper observation of student participation

5

Cormac Quigley, Elaine Leavy, Etain Kiely, and Garrett Jordan
Galway-Mayo Institute of Technology, Galway, Ireland

Introduction

This chapter shares the results and insights from a collaborative project to use learning analytics to capture and transform learning in the first year of undergraduate science programs. The multidisciplinary team is composed of academics and technical staff with a shared goal and numerous motivations. The shared goal was to use analytics to describe and optimize learning. This is an ongoing project first instigated in 2016, which has evolved from using descriptive analytics to create personalized feedback forms, to creating dashboards and is working toward using historical data to train models to monitor and predict engagement and disengagement (identify at risk students). Data are collected through a blended learning model, which has enabled students to take greater ownership of their learning and staff to enhance curriculum and learning strategies.

It is important to convey how the term blended learning strategy is being defined and conceptualized here because this affects the practice and context of this study. Blended learning combines the face-to-face interactions (lectures and laboratories) with computer-mediated instruction and assessments [1]. This offers students the opportunity to apply learning in a laboratory-based environment, while also facilitating the use of online self-paced opportunities to practice quizzes, watch videos, and complete online assignments. Virtual learning environment (VLE) is a term used to describe the online platform, which in this study is Moodle, used for the delivery of resources, activities, and assessments to learners as well as facilitating interactions.

The VLE as a source of data

Typically, educational institutions will require students to log on to individual accounts, which by necessity then create a log of the interactions each user has with the VLE. VLEs, in general, use Relational Databases to store information. The relational model lends well to storing strongly linked, transactional, and unique data such as students, course, modules, and results. Relational Databases are an established and standardized technology and are VLE agnostic. How these interactions are logged, recorded, and how the resulting data can be leveraged to enhance blended learning are explored in this chapter.

The most important consideration in producing clean data is to ensure the data fully reflect the intended course design and assessment strategy. It is also important to the project that technology carefully considers pedagogy and that the structure of the course and VLE are designed to meet the agreed educational needs and aims rather than adapting the teaching to suit the technology. This reflects literature, VLEs can offer students timely and accessible information about their progress within a course or module, but only if the VLE and module design are properly aligned [2].

In this project, this was achieved through capturing and validating blended learning data associated with face-to-face interactions in laboratories and lectures as well as the interactive activities within the VLE (e.g., quizzes and assignments). The VLE can, and should, be used to record information about student participation outside the virtual environment. Besides the fact that it facilitates the use of the data, it means that student information is held in one secure system that is maintained (often at institute level) to a commercial standard with data policy compliance, data backup, and data security to match.

Specifically looking at Moodle, the gradebook enables advanced implementation of assessment weightings, calculations, and timelines for these activities, meaning information about student progress within a course or module is readily accessible. This can be used to produce meaningful analytics that can be used directly by students to monitor progress or extracted by the analytics team for deeper analysis. Moodle has many useful plug-ins for visual descriptive analytics of components such as completion tracking, progress bars, and heat maps that offer a snapshot to both the students and/or lecturers of student participation and interactions. Again, all these are dependent on the validity of the source data to function—remaining aware of this during design of a VLE page can pay dividends.

Learning analytics in education

In today's data-driven, digital era, there is much rhetoric around data analytics; essentially the drilling, refining, and mining of data to make sense and value of these data. The year-on-year learner interactions with the VLE and other resources provide a deep source of data, which can be refined and mined to better understand a given module [3]. The term Learning Analytics defined by Long et al. offer a definition of how this study interprets the term learning analytics [4].

Learning analytics is the measurement, collection, analysis and reporting of data about learners and their contexts, for purposes of understanding and optimizing learning and the environments in which it occurs.

There are four main types of analytics, all of which can help answer different questions about learning.

- Descriptive—What is happening/happened?
- Diagnostic—Why is it happening/why did it happen?
- Predictive—What will happen?
- Prescriptive—How can we make it happen?

Graphs of the type shown in Fig. 5.1 demonstrate the range of analytic types illustrated along two axes—value and difficulty. Many industries have expanded their use of analytics to help identify the best outcomes of potential scenarios and to identify the best possible products and solutions to meet their customers' needs [5]. Similar to other industries, the Higher Education sector is turning toward analytics in the form of learning analytics to gain valuable insights into the behavior of both students and lecturers throughout their learning and teaching journeys. In turn, these insights can help make institutions make more informed, data-driven decisions with the overall aim of improving student success and, thus, meet the needs of their learners.

This linear model does not fully reflect the often-iterative interactions in higher education where insight and hindsight are gained throughout the learning process as well as at the end of academic cycles. The role of the lecturer as expert educator

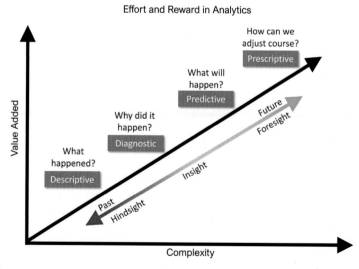

FIG. 5.1

Analytics value escalator: value versus complexity. Typically in commercial projects, the complexity of and reward from analytics follow a fixed path.

means that unlike a commercial situation where users may be trying to determine previously not understood consumer behavior, lecturers may have significant insight when presented with descriptive analytics relating to their students. Thus the relationship between value and complexity may not hold true in an educational setting, and gaining predictive and prescriptive insights may be less challenging than anticipated. The model also fails to accommodate the difference between analytics used in a purely commercial enterprise where there is a single easily defined goal (i.e., to maximize the profit produced) and an educational setting where there are multiple stakeholders, each with complex goals that can be difficult to quantitatively define and measure.

In this chapter, we explore how this understanding of what is possible through analytics and the value of feedback can lead to certain design choices in creating modules and their related spaces in the VLE.

Motivations

Several factors have driven the development of learning analytics in GMIT. The first was the automating and dashboarding of an existing feedback process that involved the creation of feedback pages that have been embedded in the School of Science and Computing since 2016 in Chemistry and Maths modules (see Chapter 9). Research conducted over 4 years revealed that frequent personalized feedback motivated learners and changed learning behaviors. It was clear that improvements to this project were, at least in part, constrained by the quality of data available. The analytics created were used by lecturers and decision makers to monitor online engagement of at risk students. They offer learners a weekly summary of their progress. Having started out as a limited project using individual plug-ins to create limited analytics, many lessons have been learned as the team has moved to creating a general dashboard system, which combines all these functions in one place.

The team was also interested in the use of data on students and their learning for enhancing the curriculum known as curriculum analytics. A quick review of descriptive analytics data horizontally and vertically can offer insights into the appropriateness and value of specific assessment for learning. Interesting findings within the study offered insight into student workload, student effort, and frequency of assessments for engagement and student success. Models generated from data patterns and student observations have challenged the widely held beliefs around the over assessment of our students and instead enabled lecturers in Chemistry and Maths to advocate for frequent and low stakes assessments to drive learning using "the currency of learners."

At an institutional level, it was hoped that the project could assist student retention issues in first year science. Scaled across the School of Science and Computing, using many data points from different modules on individual learners offers much richer and more timely data on struggling students who are at risk of dropping out.

Using learning analytics in chemistry education

It is clear that the methods and priorities of analytics development in commercial settings are not well suited to the educational environment. Clow suggests that the Learning Analytics cycle instantiates and enables reflective learning at both an individual and organisational level [6]. The cycle can be complete even where the intervention does not reach the learner who originally generated the data. Instead, if the lecturer can interpret feedback and make changes to their own resources, then the cycle is complete. Fig. 5.2 illustrates a blended learning analytics model reflective of this reality. This reforms models such as those shown with clean, meaningful data at the heart of an iterative cycle with often interchangeable transitions between the types of analytics (descriptive, diagnostic, prescriptive, and predictive).

At each analytics phase, interactions by lecturer and learner are considered. A simple example might start with curriculum design and observing the student engagement with an assessment. The premise that student participation in learning activities tends to correlate with subsequent student success can be explored. Observation from the data may highlight students who are not engaging well, which may transition the lecturer into the diagnostic phase to explore why this might be happening. An intervention informed by consultative interaction with the learner may result in a curriculum intervention such as extended deadlines. Curriculum or learner interventions may be prescribed for the collective student group throughout or at the end of a learning cycle. Similarly, the curriculum or hidden curriculum (in terms of social

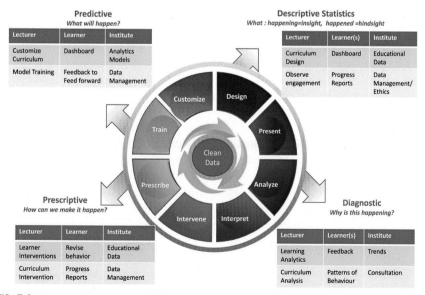

FIG. 5.2

Learning analytics. Learning analytics is distinct in nature to other commercial applications of analytics because at each step, there are multiple stakeholders.

influences) may be customized (a prescriptive change) to attempt to alter learner behavior and change predicted outcome.

Implementing predictive analytics comes at many levels. It may be a lecturer-driven insight based on descriptive data, or it may be based on automated predictive models. The analytics features in newer versions of Moodle (3.7 onward) now allows the use of historical data to train models that may be used to alert staff of at-risk students. This may be very useful to build intelligence into the VLE environment and reduce time for lecturers in the diagnostic phase. It also unlocks the true value of predictive analytics, an ability to gain otherwise hidden foresight using data analysis.

Learning Analytics is an evolving, multidisciplinary research field which requires researchers with a mix of skills from several disciplines. It is critical that the science of learning and curriculum design is kept at the heart of the analytics solution, with clean data reflective of the need for the data to be representative of the pedagogy and practice. Lecturers and students must remain as key stakeholders in the design and implementation of Learning Analytics [7]. For this reason, with clean data at its core, the model identifies the key interactions/outputs by lecturer and learner(s) at each of the analytics stages. Early engagement with your relevant Technical Department may be necessary to ensure solutions are technically possible and help enable the future design, build, and implementation of data collection in course design. It has been the experience of the project team that having clean underlying data makes any steps into analytics much more likely to succeed.

Choosing a starting point

The most common starting point to any data analytics project is to employ descriptive analytics—this will help you explore, understand, and "feel" your data and see the data come to life with data visualizations such as graphs and charts. It is important to remember that Learning Analytics is not just predictive analytics, and much can be achieved with Descriptive Analytics such as the creation of dashboards. In fact, the present focus of our team is to enable extracting data to Power BI for dashboard design, allowing access to descriptive learning analytics for any lecturer who is so inclined. Not to be underestimated, descriptive analytics can represent the perfect starting point for an analytics project and will help pave the way should you wish to progress to other types of analytics. Learner Dashboards can benefit both staff and students without the huge time and resource overheads required to build and implement a more complex predictive analytics model and, in many cases, can allow staff and students to gain their own diagnostic and predictive insights. This is, of course, contingent on suitable data being recorded within the VLE.

At the heart of every data analytics project is the data itself. Data are the solution but also can be the problem. The meaning of this last sentence relates to data quality. Most likely, volumes of data have been generated, stored, and are available to tap in to. However, what is important is data quality. Is there missing data, incomplete data, mislabeled data, outdated data, or duplicate records? All these contribute to the presence of "dirty" or "bad" data, which ultimately can lead to failure and/or delays in analytics projects.

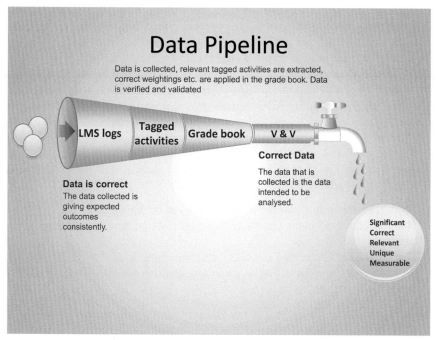

FIG. 5.3

The clean data pipeline. As data moves toward an intended output it is constantly refined and validated.

The team has examined the data flow through our systems based around Moodle illustrated in Fig. 5.3. This signposts how data enters, where it is stored, and how it looks when extracted in raw format. The less cleansing and processing required, the better—one to remember is Garbage In Garbage Out! A recommendation that cannot be overemphasized is to ensure the learning strategy and gradebook is setup correctly from the start within the VLE.

Poor-quality data costs time and resources and if presented to users (lecturer or student) is likely to cause disengagement with, or distrust of, the learning analytics. This can prevent further uptake of analytics and reduce the value of future learning analytics outputs, so the main objective is prevention rather than cure. Prevention measures can be employed to help avoid accumulation of bad data such as discussion and agreements about data standards for the labeling and entry of data, user training, and periodic data checks. It is worth trying to stop the pollution of data sources sooner rather than later. The longer it is an issue, the more difficult it becomes to clean up data. Where findings are dependent on data input for accurate results, it is essential for accurate data to be entered.

Our experience with creating and using clean data

In our case, student participation data were monitored specifically regarding engagement and disengagement by the analytics team. The overarching guiding question was to find out whether the data offered a true descriptive presentation of student engagement or disengagement (i.e., at risk of dropping out). The data were validated at intermittent periods for accuracy and appropriateness. The following descriptive metrics were identified in the blended learning experience: Learner Presence (attendance in lectures, tutorials, and laboratory practicals); Learner Engagement/ Disengagement (heat map and progress bar completion tracking, time since last logon to VLE); and Learner Achievement (grades in formative and summative assessments) Learners' data were captured and presented to the students within the VLE gradebook environment. In many cases, simple steps can ensure many problems can be avoided. For instance, when recording learner attendance using the VLE, it is important that this is recorded reliably and in the same manner by all staff. Some guidelines for attendance recording are as follows:

- There are not multiple places that attendance can be registered
- That it is registered in the same way by all staff—(e.g., how do staff treat latecomers?)
- That attendance is recorded correctly (students can be invited to check); this is particularly relevant for large classroom settings where students may need to check-in themselves
- That records are kept up to date.

Other learner engagement such as Moodle Heatmap plugin provides a visual representation of the number of users who have accessed a resource (unique viewers) and the total number of times the resource has been accessed. Again, this is very easily implemented but is contingent on access to the resource being available only through Moodle. For example, sending a resource to the class by email as well as placing it within the VLE will mean that accessing the resource will not be recorded for many users (or at least not in a useable way). Instead, sending a link to the VLE will ensure that useful data are recorded. This may also be important if you want to use analytics services based on third-party services such as YouTube or Stream. By restricting access to resources only through defined links ensures that the data used for analytics are representative.

At a single-item level, setting up assignments and resources with completion tracking and dates allows for students (and staff) to be given a visual insight into their individual progress through the learning material available to them. The completion progress tracker in Moodle enables this visual insight to be delivered seamlessly within Moodle but can only operate with the data recorded within Moodle. Again, even where assessments are completed offline, grades can be inputted into the VLE to enable this visualization but with the same caveats as applied to attendance recording—accurate, up to date, and correct.

At a more advanced level, a built-in configurable reports function within Moodle allows lecturers to create custom reports with very limited knowledge of SQL. The configurable reports plug-in allows for creation of reports drawing directly on the data tables within Moodle without any need for additional download or processing. The range of reporting is extensive. Useful examples for student monitoring include reports such as lists of learners who did not log on in the past 7 days or lists of users who have completed particular courses, whereas at a curriculum level, reports can monitor usage of different activities, activity types, or simply number of actions per course in a given time frame. Although this has the potential to allow powerful analytics, it is limited by the fact that it runs queries on the live database. This should be avoided in practice because it can lead to instability. Instead, a more sensible approach is to run queries on a duplicate, offline version of the database. This duplication has many possible forms, so it is likely that an interested lecturer will need to liaise with technical support to be able to access this possibility. Again, however, the emphasis of this chapter is that, for any of this to provide meaningful output, the database must contain clean data.

Findings and conclusions
Designing blended learning for clean data

Thought needs to be given to ensuring that data are entered/recorded in a manner that ensures, as far as possible, it is significant, measurable, useful, and reliable. To achieve this, things to be considered include:

- Having a consistent naming scheme for all activities
- Correct VLE data setup, for example, gradebook, weightings, timelines, etc.
- Labeling of data entries
- Identify key interactions (what should be measured)—quizzes, assignments, forum posts, attendance, and interactions with peers

None of these factors should alter the intended course design—it is as easy to set a course up correctly as incorrectly if sufficient forethought is given. If there is an intent to complete analysis over more than 1 year, then it is important that these are also consistent from year to year (or iteration to iteration of a module). Having data entries labeled consistently in different iterations of a course enables easy comparison over a longitudinal study. Making substantial changes to items between years without a change of labels will render subsequent year-to-year analysis, at best, useless.

Evaluation/validation phase: Does the data reflect reality?

Our data consumers (students and staff) have little patience for poorly represented and incorrect data. Throughout this project, nearly all negative feedback from students regarding analytics has involved incorrect data. It is crucial to ensure that quality assurance processes are well considered, planned, and implemented. This is essential for reliable student facing analytics.

In predictive analytics, this may involve the validation and verification of model accuracy and performance by manually assessing outputs. This, of course, can and should be done with the help of students—as long as they are aware they are helping to develop a new system and the outputs they are seeing are not validated. This will help provide insights into how and where predictive models need to be tweaked.

For descriptive analytics, the process will be slightly different. What needs to be examined is how correct your data representations are and if your visualizations are looking and behaving as expected. It is advisable to take it step by step, tracing the data right through from very beginning to the very end of the process. Look at some example records in your starting dataset and follow through to the end point of chart, graph, or table to ensure correctness of data. Is it correct and does it reflect reality? Again, errors are often quickly discovered by students, but in this case, they should be avoidable.

The motivation behind these checks is to prevent errors that creep into transformations or aggregations of data, particularly if there are elements of manual effort involved in pulling in, transforming, or setting up automated processes. Manual manipulation of data can very easily introduce an error in the automated process, and the computer is just a dumb terminal that will do as it is told!

- Start with class lists—are they all are your students and are all your students there? (do I need an enrollment key for my course?)
- Think of possible exceptions and how they should be handled/catered for:
- Examples may include repeat students and students with exemptions—will they skew statistics?—do they need to be excluded or catered for as exceptions?
- A quick and simple test plan can be beneficial:
- Draw out a table listing test cases specifying inputs, expected results, and actual results.
- Complete your testing, test case by case, and observe differences between expected and actual results—understand why and note for later discussion and fixing.
- Think about how you wish and expect data to be presented. Is it presented that way, if not, why?

What will it answer?

The data stored in your VLE can answer a range of questions. As with any set of data, it is possible to generate descriptive statistics to give you an overview. In the case of the VLE, this can be used to give a visual description of individual performance, class performance, and performance in individual course components by way of grades, attendance, time of access, and frequency of access. Whether or not this is of use will depend on how accurately the data is stored and how it is interpreted, leading to the emphasis on quality of data rather than quantity. As the quantity of data builds up, it is also possible to assess correlations between different factors. This can be done manually or automatically and can identify unforeseen factors affecting student performance.

More important than this might be what it cannot answer. It is worth recognizing that although we are always trying to measure student engagement, it is very difficult to measure this directly. It is important to set expectations for the target audience in this regard. An example from our project was analysis of recording of time on task using different built-in measures and custom plugins designed to capture mouse activity was performed. Analysis of data from classroom-based cohorts of students who were completing a credit-bearing assessment showed that none of the measurements accurately described how students had engaged with the task. Recorded interaction times varied by a factor of 6 and the measure closest to the lecturers' estimation of student time spent (calculated based on start and finish times for the examination) was still a significant underestimate. Even where it is possible to capture time spent by students on given tasks, there is no way to evaluate the quality of their engagement. Rather than measuring this directly, proxy measurements can be used such as time between sequential tasks.

What can be learned for students?

For students, there are obvious advantages to having a blended learning environment that returns clean data and analytics. They are empowered with feedback on their own performance, making goal setting and performance expectations more transparent. The experience of the project team has been that students, given transparent feedback on their own performance, are likely to adjust their behavior to improve their own performance. The knowledge gained by students through clean and accessible data is relevant only for the duration of the course, and although it may have long-term impacts on the students' development, it does not in itself promote long-term changes in teaching practices.

What can be learned for lecturers?

Lecturers, on the other hand, are often assigned to teach courses for many years, so the data can be used to drive short- and long-term feedback cycles. For lecturers, the examination of student engagement and performance can give insight into what students find difficult, what resources students spend time engaged with, and other factors that might affect student performance. This can take place at curriculum level, at individual resource level, or even at the level of individual assessment questions.

Curriculum-level analytics

An example of curriculum-level analytics from practice sought to categorize student perception of the workload based on how they interact with the VLE. Procrastination is pervasive among undergraduate students [8] and is a result of multiple factors. Observation of student interactions can provide insight into this phenomenon by evaluating whether deadlines set are correlated to the time at which students complete the required activity indicates whether or not procrastination is taking place.

An observation of this phenomenon showed that for low-stakes routine activities such as weekly quizzes, very few students procrastinated. Quiz deadlines were set far in the future (up to 2 weeks), but students would, on average, complete them several days ahead of the deadline. In addition, where the quiz deadlines were set at differing distances from the quiz opening time, the average time before deadline which the quiz was completed were statistically independent ($P < 0.001$).

This could be contrasted with nonroutine quiz assessments where students were required to complete a summative quiz, which was not part of a weekly routine work and had a greater weighting (although still only approximately 1% of the overall module grade). The students could be observed to procrastinate. Almost 50% of all quiz attempts (181/366) were made on the day of or immediately before the deadline despite the deadline having been set 2 weeks in advance.

What enables this analysis is the presence of reliable data on student interactions. The data in this case required cleaning, such as removal of nonattempts (where a student accesses the quiz but makes no meaningful attempt to complete it). In addition, the setting of time limits on quiz attempts also prevents students completing quizzes over multiple days, which would render the analysis above much more difficult.

A second example of the use of descriptive analytics was that students who watched prelaboratory videos were better prepared for both weekly laboratory practicals and the final practical assessment ($n = 242$, $P < 0.0005$). This insight was only possible through the use of analytics in combination with the use of surveys; however, it would be impossible were the VLE not set up to control students' access to specific resources and provide cleanable data. These observations offer something to think about course design and, of course, naturally lead to timeless debates on the nature of higher education and the need for learner independence. Notwithstanding this, such observations leave the lecturer armed to make informed decisions about course design and how they can affect student interactions with an assessment through course design. Such information also leaves a lecturer armed with factual information to challenge assertions such as "overassessment" with clear and definitive analysis. These two observations should make clear that the creation of descriptive analytics naturally prompts the lecturer to make diagnostic, prescriptive, and predictive inferences about their own teaching.

Assessment-level analytics

It is also possible to use analytics at a much more detailed level to look at interactions with individual assessments, questions, and resources. The data gathered during online assessment can provide a rich source of insight into how students perform in each aspect of the assessment. Reports of student results for a particular quiz can offer insight into student learning such as common mistakes, relative difficulty of questions, or even at a simple level, identify incorrect or poorly written questions. Although in theory, each student response can be taken from the database for analysis, in practice, with some forethought, a large amount of information is readily accessible if an assessment is well designed.

Moodle itself contains inbuilt analytics that provide descriptive statistics on all quiz results. At a quiz level, it offers basic insights such as mean, median, and standard deviation of results. This provides immediate oversight of whether a quiz is performing as intended. It also offers similar basic statistics on each question that can highlight questions that may have errors or a significantly higher or lower degree of difficulty than intended. There are also more advanced descriptive statistics provided showing the performance of students in individual questions compared with the quiz as a whole and the relative contribution of each question to students results. These can be used to assess the relative difficulty of particular questions in comparison with the entire quiz and whether there is consistency in which questions prove challenging.

Again, whether the lecturer intends to have each question be of equal difficulty or that the entire quiz would be an assessment of the same skill is a matter of curriculum design but having the descriptive analytics available means that lecturers can make informed decisions. A note from practice; a quiz in which the order of the questions is completely random is far less amenable to analysis than a quiz in which all students take the quiz in the same order. Moodle will provide analysis for each question and each random variation of a question.

In practice, this feedback has enabled improvements to assessments. It provides simple error checking, identifying questions with errors, and either fixing them (where the question is incorrectly marking) or removing them (where the question itself is incorrect). It is almost unavoidable that errors will occasionally occur in questions, but it is possible to change the correct response and regrade the question after the examination. This is particularly important in summative assessments where the marking is not made available to the student after the quiz. It also informs future design such as reordering quiz questions to place harder questions at the end of a quiz and removing questions of dissimilar difficulty from a pool or random questions. These allow data-driven curriculum-level improvements for future iterations of a learning resource or assessment.

Where are the pitfalls?
Of course, the examples just described may raise as many questions as they answer. In all situations, when attempting to use data to answer questions with a view to changing future behavior, it is imperative to remind oneself of Goodhart's law. Although widely known, it bears repeating, "When a measure becomes a target, it ceases to be a good measure." This maxim is entirely applicable to the use of data in assessing higher education. A typical example might be that analysis of student results shows that those who access the course more frequently obtain, on average, higher grades. Although this may be true, were the lecturer to decide that the frequency with which students access the course was then a target to be met, it is likely that some students would access the course only to meet this target rather than to engage in learning. In fact, this might even reduce the achievement of learning outcomes because it would distract from time spent on other learning.

This example then is an ideal example of what is not included in the data. In many instances, it is possible to show correlations with a very high level of statistical significance between different variables of student behavior. "The data can tell anything. You have to ask the right question."—Unknown. Rather than suggesting the answers to all issues around student engagement are to be found in the data, this suggests that, with the right manipulation, you can make the data say whatever you want it to.

In another example from practice, once again about time on task, it was found that the students who spent more time completing a quiz would perform less well. (Sample size of 2125 quiz attempts across 7 low stakes quizzes, correlation coefficient $= -0.309$, $P < 0.0005$.) Here then, it is clear that the data captured are limited in what they can tell us. A naïve attempt to improve this situation would be to attempt to encourage students to spend less time on the quiz in the hope that this would improve the grades achieved. It seems obvious that this is unlikely to work because the correlation is most likely because of the different levels of preparedness of students and giving poorly prepared students less time to complete the quiz is unlikely to improve their grades.

There are also ethical pitfalls that arise with the collection of data. Aside from the obvious ethical and legal considerations, many of which are dealt with by data protection legislation, there can also be unintended consequences. Many institutes also have policies surrounding the use of data for evaluation of staff either restricting it to management or human resources or banning it outright. On the other hand, other resources are not subject to these protections. Before performing analysis, be careful that you are willing to accept the results. For instance, if analysis shows that students who draw the short straw and are timetabled for later classes are significantly more likely to drop out, is it ethical to continue this practice, or will you need to offer some mitigation for this? If certain facilities are consistently correlated with poorer student outcomes will you be able to accept this and make changes needed?

An eye toward what is next

In our own projects to date, we have used what we have access to and embraced the power of Moodle's configurable reports along with logging and reports functionality on Moodle live data. This has worked for the purpose of producing descriptive analytics, which was the first stage in our learning analytics project. This solution, however, is not sustainable or scalable enough to bring us to the next level of learning analytics. We, unsurprisingly, wish to progress forward to using predictive analytics and machine learning on big data. This is a predictable situation and we are well set to make this step because we have a foundation of clean data on which to build.

No matter what avenue your learning analytics path may take, one of the limiting factors on its usefulness will be the data that it can make use of. Having spent considerable time investigating Moodle Analytics to see how well a standard "out of the box" Moodle analytics solution meets our needs and is fit for purpose. The plugin allows us to create and train models using current data to move beyond descriptive analytics to include predictive, diagnostic, and prescriptive analytics to inform and enrich learning in GMIT. The biggest limiting factor in the use of this and similar systems is the lack of clean data within the VLE. The constantly changing of learning environment in response to COVID 19 has made this all the more challenging. All the while, we are moving toward designing and building our own in-house predictive analytics system that will integrate other data sources such as Office 365.

What is clear to date is that no matter the eventual destination of analytics, it will be best served by having solid foundations in clean data.

To use produce clean data for descriptive, diagnostic, prescriptive, and predictive analytics effectively, the team continue to work on the following:

- Developing and maintaining a standard operating procedure (SOP) regarding standardization of data-naming conventions, data labels, protocols, and intended outputs. Ensure people are on the same page and understand the need and importance of data quality.
- Ensure access to a live Moodle database on a server or data warehouse, which is completely isolated from the production environment to use for data analytics. The institution is critically dependent on a stable live VLE environment. Any project cannot jeopardize the VLE by running queries/reports that might slow down or crash the system.
- Hive off/store annual data—do not lose/waste your valuable asset. Every year, there is a new Moodle instance and database created. Back-ups of old databases instances can be requested but take time to request, receive, and get stored in insolation. Moving forward, these data need to be part of one central store. Data mining and machine learning depend on big data so that algorithms can learn effectively. Hiving off of annual data will help achieve that goal.
- Related to above point—Data Lakehouse is the new data management paradigm that combines the capabilities and benefits of data lakes and data warehouses. Building a Data Lakehouse will enable benefits such as—Real Time Analytics, Data Science and machine Learning, Reporting and BI—all in a scalable and secure manner. The development of such a Data Lakehouse can cater for both structured and unstructured data, which will help futureproof for new types of future data. This can accommodate many disparate systems such as student registration system, virtual learning systems, and timetabling systems, essentially bringing together data sitting in separate isolated silos. Bearing this in mind at the design stage as new systems are introduced will also allow for analytics to become more useful.

The success in this project has been the multidisciplinary team bringing diverse skills and perspectives. The team has worked within a limited budget relying on the VLE and Microsoft tools for collection, extraction, and analysis of data. The ground-up approach emerged from a collaboration of individuals with a genuine interest in developing analytics to enrich learning within the institute. The combination of technical skills to enable plugins and run SQL queries merged with real-time access to chemistry learner's data has offered valuable learning. It is hoped that this chapter will aid others who wish to undertake similar initiatives. The key message the team would like to share is the importance of designing blended learning opportunities to ensure you are producing clean, meaningful, and useful data. Once the data exist, they open the door to empowering students and staff with the observations needed to drive effective learning.

References

[1] Bonk CJ, Graham CR, Cross J, Moore MG. The handbook of blended learning: global perspectives, local designs. Wiley; 2012.

[2] Anon. The SAGE handbook of research on teacher education; 2017. https://doi.org/10.4135/9781529716627.

[3] Herodotou C, Rienties B, Boroowa A, Zdrahal Z, Hlosta M. A large-scale implementation of predictive learning analytics in higher education: the teachers' role and perspective. Educ Technol Res Dev 2019;67:1273–306. https://doi.org/10.1007/s11423-019-09685-0.

[4] Lockyer L, Dawson S. Proceedings of the 1st international conference on learning analytics and knowledge. New York, NY, USA: Association for Computing Machinery; 2011. https://doi.org/10.1145/2090116.2090140}.

[5] LaValle S, Lesser E, Shockley R, Hopkins MS, Kruschwitz N. Big data, analytics and the path from insights to value. MIT Sloan Manag Rev 2010;52:21–32.

[6] Clow D. The learning analytics cycle: closing the loop effectively. In: LAK12: 2nd international conference on learning analytics & knowledge; 2012.

[7] Tsai Y-S, Rates D, Moreno-Marcos PM, Muñoz-Merino PJ, Jivet I, Scheffel M, et al. Learning analytics in European higher education—trends and barriers. Comput Educ 2020;155:103933. https://doi.org/10.1016/j.compedu.2020.103933.

[8] Steel P. The nature of procrastination: a meta-analytic and theoretical review of quintessential self-regulatory failure. Psychol Bull 2007;133:65–94. https://doi.org/10.1037/0033-2909.133.1.65.

Adopting a flipped classroom to teach and learn SciFinder in an undergraduate chemistry laboratory course

6

Hafiz Anuar[a], Yongbeom Kim[b], Tag Han Tan[c], and Fun Man Fung[a,d]

[a]*Department of Chemistry, National University of Singapore, Singapore, Singapore*
[b]*School of Computing, National University of Singapore, Singapore, Singapore*
[c]*NUS High School of Math and Science, Singapore, Singapore*
[d]*Institute for Applied Learning Sciences and Educational Technology (ALSET), NUS, Singapore, Singapore*

Background

SciFinder is a powerful search engine by the Chemical Abstract Services (CAS) that fetches high-quality and relevant chemical articles from its comprehensive database in a short span of time. Widely used as a scientific research tool for many disciplines such as Chemistry, SciFinder is able to provide easy access to chemical literatures, substances, and reactions [1]. The encyclopedic nature of SciFinder allows the users to access the information of more than 126 million chemical substances with a unique CAS Registry Number tagged to individual substances [2]. With a clean and user-friendly interface, SciFinder is designed to allow its users to conduct their literature search for chemical abstract and information simply by searching for the topic, author, and/or substances [3]. Furthermore, SciFinder also has the capability to narrow down search to provide a more accurate and relevant search using additional keywords, time period, and other indicators [4]. Besides the ability to perform literature search using the name of the chemical or the reaction, SciFinder can also search for specific complex chemical structures, substructures, or reactions without the need to know the name of the chemical substance or reaction. Users can perform the search by drawing the molecular structure or the reaction using the "Drawing Editor" function. With such capabilities as a search engine, SciFinder enables University students to perform search for literature data and journals from more trustable sources instead of obtaining their results from Google or other search engines, which may not be able to provide them with the most accurate or suitable answers that they need. This is especially important for the third- and fourth-year

Technology-Enabled Blended Learning Experiences for Chemistry Education and Outreach.
https://doi.org/10.1016/B978-0-12-822879-1.00010-X

Undergraduate students, particularly when they are writing reports for their research projects, such as during their Final Year Projects. Hence, there is a need to equip the students with the skill to perform searches using professional search engines such as SciFinder. Furthermore, the ability to perform a quick search for relevant information becomes pertinent when University students are time-starved, juggling many projects, assignments, and lectures with the various modules that they are reading. SciFinder, therefore, plays an important role in effectively reducing the time needed to search for relevant information when compared with sieving through millions of articles that might not be necessarily relevant on the other general search engines. Hence in our study, SciFinder has been incorporated into an undergraduate chemistry laboratory courses, CM3291—Advanced Experiments in Organic Chemistry and Inorganic Chemistry at the National University of Singapore. The flipped classroom approach was then implemented to teach SciFinder for 51 third-year students reading the module. As some challenges have been identified in existing methods of introducing SciFinder to students, and hence, we adopted the flipped classroom approach in the study to address the issues of the current approaches [5, 6].

Purpose of this project

Currently, institutions that subscribed to SciFinder provide either online or physical training sessions for students [3, 7–10]. However, students find it hard to learn SciFinder to obtain appropriate contents for their assignments, when taught through either method. With only online lectures, students are unable to clarify their doubts after watching the online videos. Furthermore, they are unable to practice with hands-on session to familiarize themselves with SciFinder. With solely the "Face-to-Face" (F2F) session, there is limited time to interact with the students because the focus lies on teaching the actual content of the modules [5, 9]. The flipped classroom approach comprises of the online lecture and an F2F session. In the flipped classroom approach, the librarians will prepare and upload the lecture contents online before the F2F session. Students will then be able to watch and learn the contents before the class at their own pace and at own time and own target. The purpose of having a flipped classroom approach is to ensure that no student is left behind in terms of content [11, 12]. During the F2F session, instructors enhance students' learning by conducting well-designed exercises to enhance students' critical thinking regarding the topic and building on top of the knowledge that they have acquired through the online lectures. Hence, it would allow the students to master the lecture contents faster and more efficiently as compared to solely the online lectures or the physical session as the students came prepared with some previous knowledge gained after watching the online lectures [13].

Previous methods of teaching SciFinder

Various institutions are currently teaching SciFinder to their students, highlighting the wipespread use of SciFinder in the scientific community [2, 10, 14–24]. Although most institutions have used either the online tutorials or the F2F approach, no institution have yet to use the flipped classroom approach in teaching SciFinder. Rider University used Adobe Captivate 4 software to create an online tutorial for SciFinder. Upon completing the online tutorial, the students were required to write a research report about a selected named reaction with SciFinder suggested as the principal tool for this purpose. After submitting the first draft of the online tutorial, an anonymous quiz was then administrated to evaluate students skills in using SciFinder [8]. Serving as a good example for the F2F approach, The University of Colorado Denver Auraria Library uses a library instruction section. Beginning with some general information about SciFinder, the students then complete a short preinstruction survey. Using LibGuide, the librarian then highlights and demonstrates each feature in SciFinder and for the remaining of class time, the students will complete a printed exercise related to SciFinder. The exercises are nongraded and anonymous so as to ensure minimal stress. Similar to the previous example, SciFinder was previously taught to the Undergraduate students at NUS (National University of Singapore) through a one-time F2F library training session, where students came without any previous knowledge of using SciFinder. With this 2-h library training session as a tutorial on how to "use SciFinder," students learned the functions of SciFinder, such as obtaining journals using SciFinder via reaction structure, which is relevant to the module. In the 2-h session, students also experienced a hands-on practice session through a sample problem involving a mini assessment to find a relevant journal using the shortest pathway possible. The usefulness of this session was quantified by the grade given to the quiz based on the accuracy and number of steps taken to retrieve the journal. The greater the accuracy of the journal and the fewer the number of steps taken, the higher the score attained, implying the effectiveness of the session. However, it was found that there were several limitations to teaching SciFinder using solely the F2F session. Because the F2F session had to be accommodated and fitted into the lesson time, the lesson was merely a short 2-h session. Given the complex and profound nature of SciFinder, the librarians had to cover the contents slowly because it was the students' first encounter with the program. Hence, there was only a limited range of content that the librarians could cover. As a result, achieving deeper learning and metacognition was more difficult with just one instance of encountering the content and program without any prior preparation. Furthermore, because there were no follow-up lectures, students found it difficult to process bulk amount of information of how to use SciFinder quickly, given the lack of other opportunities to understand and practice.

Alternative approach: Teaching SciFinder through flipped classroom

The flipped classroom approach was designed to solve the limitations of teaching SciFinder to students using solely physical sessions and to maximize time utilization for physical sessions [12, 13, 25].

Previously, there was little prework, and the contents were transmitted to the students during the 2-h physical lecture, meaning there was a limit to the amount of knowledge transmitted and absorbed. Then, most of the learning happened on their own when they used SciFinder, for example, when they did the assignment itself. The amount of knowledge gained by the student was represented by the size of the rectangular box as seen in Fig. 6.1.

In the last semester, the flipped classroom approach was adopted to enhance students' learning and to reduce the time taken for each library session from 120 min to 90 min. Although this will not affect the amount of content that is being taught to the students, students' in-class time will, however, be maximized. Currently, SciFinder is taught by a trained expert in the 90-min library session for 25 students per run. Before the library session, the students were given access to SciFinder registration and introductory online lectures catered for preclass preparation. With the access to the introductory lectures online, the students can now watch the videos at their own pace before attending the lectures. The introductory lectures now give them an overview of what they can expect in the F2F session. They can also try hands-on prior to the F2F session and clarify any doubts regarding the use of SciFinder during the F2F session. The students now have a longer time to process the bulk information and even try hands-on before going for the F2F session. As it will not be the first time that

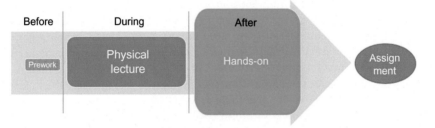

FIG. 6.1

Learning process before teaching SciFinder using "flipped classroom."

Courtesy Fun Man Fung, Yongbeom Kim, Tag Han Tan.

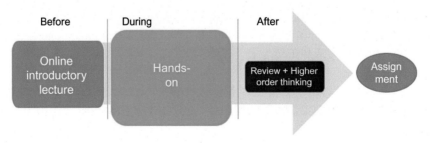

FIG. 6.2

Learning process after teaching SciFinder using "Flipped classroom."

Courtesy Fun Man Fung, Yongbeom Kim, Tag Han Tan, National University of Singapore.

the students are seeing the contents and the students now have a better idea of what they are learning, the librarians now can go through the contents faster, giving them more time for hands-on experience and practice to enhance their learning. As a result of this flipped classroom approach, the in-class time (F2F session) is maximized with a better learning experience for the students.

With the use of the flipped classroom, a large part of the learning now takes place even before the physical lecture, as shown in Fig. 6.2. Now, during the physical lecture, they are able to experience SciFinder hands on and the flipped approach of teaching SciFinder. This allows for a review of the materials that were covered when they are attempting the assignment after the physical lecture and, hence, develop a higher-order thinking when compared with what was shown in Fig. 6.1.

Methodology

In this flipped classroom approach, the teaching involved two stages and two groups of personnel involved, namely the lecturers who will work together with the librarian to craft the course design and the librarians who will prepare the materials and conduct the F2F session.

Stage 1: Preparation for Face-to-Face (F2F) session (course design)

This stage primarily involved the crafting of the course design by the lecturers and librarians, including how to apply what they have learnt from the session to their assignments in the Undergraduate Chemistry Laboratory Course. The librarian will then begin to prepare of the contents of the physical session, including the in-class activity, and the uploading of the introductory video [26] on SciFinder, 2 weeks before the hands-on library session. The online introductory video gives students a basic concept of how to use SciFinder, which covers the different functions of SciFinder and the use of a function known as *"Reaction Structure"* in SciFinder. As this session was conducted for an undergraduate chemistry laboratory course, the SciFinder introductory video focused on reaction searching using the *"Reaction Structure"* function. This function required students to perform a search for journal literature based on a drawn reaction scheme, as shown in Fig. 6.3. This method of searching will subsequently be assessed via the mini quiz. Students were reminded during lectures and via emails to view the video content before the F2F session.

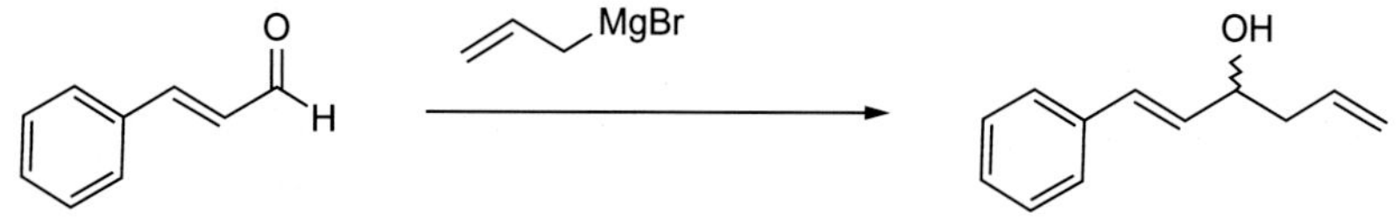

FIG. 6.3

Grignard reaction of trans-cinnamaldehyde. Example of a reaction scheme that students can draw using SciFinder to search for relevant literature.

| Recap & quick demo | Hands-on practice | Assessment | Application of knowledge |

FIG. 6.4

Flow of the F2F session.

Courtesy Fun Man Fung, National University of Singapore, Singapore, Singapore.

Stage 2: Conducting the F2F session

During the actual F2F hands-on session, the librarian will conduct the session and guide the students in retrieving the information using SciFinder. As seen from Fig. 6.4, the librarian will first give a quick recap and demonstration on how to use SciFinder, as the students should already come prepared by watching the introductory online video before the F2F session. The quick demonstration will cover mostly reaction searching using the *"Reaction Structure"* function in SciFinder and the instructions of navigating in SciFinder. Students were then given time to try some hands-on practice and clarify their doubts before attempting the assessment given to them. Within the 90-min session, students were tasked to complete a mini assessment that mostly included the flipped classroom *("Reaction Structure" search)* contents. The assessment was graded subsequently. After the session, students can then use the skills and knowledge taught in the flipped classroom in their reports. With this flipped classroom approach, students do not have to wait for the F2F session for SciFinder to be taught. Students can now watch the videos and attempt to use SciFinder on their own prior to the session. This would mean that students will have at least an overview or a background knowledge prior to attending the session, implying that the librarians can hence conduct the class at a faster pace in tandem with using higher-order thinking exercises in enhancing students' learning. They can then review the contents at home after the session, forming a continuous learning process.

Results and discussion

The grades attained for the in-class assessment of the students under the flipped classroom approach were compared to the grades of the students of the previous semester. As seen from Table 6.1 and Fig. 6.5, there was a noteworthy improvement from the previous semester where only 49% of the cohort received an "A" grade compared with the 77% of the students who have obtained an "A" grade in this semester under the flipped classroom approach. Generally, the students have fared better in their assessments in Semester 3 after watching the online video and going through the tailored hands-on session. This is backed by the notable rise from 49% to 77% who

Table 6.1 Tabulated grades attained for in-class assessment in Semester 2 (nonflipped) and Semester 3 (flipped).

Grade attained for the in-class assessment (%)	Semester 2: Nonflipped	Semester 3: Flipped
A	49.1	76.5
B	44.4	21.6
C	6.5	2.0
Total	100.0	100.0

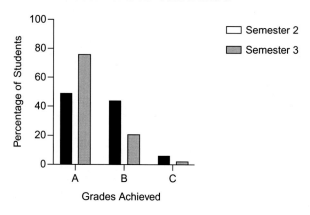

FIG. 6.5

Comparison of grades attained for in-class assessment for Semesters 2 and 3.

Courtesy Tag Han Tan, Yongbeom Kim, NUS High School of Mathematics and Science.

received "A" as mentioned earlier and a dip from 6.5% to 2.0% who received the "C" grade. Based on the results alone, there is already a strong indication on how teaching SciFinder through the flipped classroom approach has positively impacted the students. While the results alone gave us a quantifiable indication of the success of the flipped classroom approach, the success of the flipped classroom approach was also indicated through the learners' perception, shown by the improvement from the pre- and postlibrary session survey. As seen in Table 6.2 and Table 6.3, in the class of 51 students (Semester 3), 43 and 50 students responded to an anonymous perception pre- and postlibrary session survey, respectively, regarding the flipped classroom approach *(response rate: 84% and 98%, respectively)*.

In the postlibrary session survey (Table 6.3), 94% *(strongly agree and agree)* of the cohort indicated their familiarity with the tools to conduct literature and finding relevant journals with SciFinder. This is indeed a significant improvement in comparison with the prelibrary session survey, where only 19% were familiar with the

Table 6.2 Learners' perceptions of the flipped approach system (prelibrary session).

Survey statement for student response	5	4	3	2	1	5+4	2+1
I am familiar with the tools to conduct literature search using SciFinder	2	6	8	17	10	19%	63%

The scores from 5 to 1 represent the following agreement levels: "strongly agree," "agree," "neutral," "disagree," and "strongly disagree," respectively. The total number of responses for each level of agreement is tabulated. The combined category "5+4" represents the percent of students responding with "strongly agree" and "agree"; the combined category "2+1" represents the percent of students responding with "strongly disagree" and "disagree." N=43 students.

Table 6.3 Learners' perceptions of the flipped approach system (postlibrary session).

Survey statements for student responses	5	4	3	2	1	5+4	2+1
I am familiar with the tools to conduct literature search using SciFinder	12	35	3	0	0	94	0
The SciFinder video guide helps me understand and complements the F2F library session	8	34	8	0	0	84	0

The scores from 5 to 1 represent the following agreement levels: "strongly agree," "agree," "neutral," "disagree," and "strongly disagree," respectively. The total number of responses for each level of agreement is tabulated. The combined category "5+4" represents the percent of students responding with "strongly agree" and "agree"; the combined category "2+1" represents the percent of students responding with "strongly disagree" and "disagree." N=50 students.

tools in performing the aforementioned steps (Table 6.2). More importantly, 84% of the students answered that the SciFinder video [27] uploaded online helped them in their learning and complemented the F2F library session (Table 6.3). Because the videos [26] are easily available online, the students are able to access the videos at any time and learn the content at their own pace, allowing them to digest the content extensively. Students would then be more prepared when entering the tailored F2F session. The F2F library session then assists students to integrate the entire learning package as a whole.

Table 6.4 Tabulated results from a postlibrary session survey question.

Which approach would you think you learn better?	Choose one option only
The F2F library hands-on session AND online video	29 (58%)
The F2F library hands-on session ONLY	14 (28%)
The online video ONLY	2 (4%)
No preference	5 (10%)

As seen in Table 6.4, 58% of the respondents indicated a preference for the flipped classroom approach. By contrast, 28% indicated a preference for only the F2F session, whereas a mere 4% indicated a preference for only the online video. This suggests that both approaches alone are insufficient to teach students and a flipped classroom approach must be incorporated in making the learning session more purposeful. Hence, we can conclude that the flipped classroom approach is the most favorable and effective approach in learning how to search information from databases such as SciFinder [24, 28–30].

Limitations and further improvements

While the flipped classroom approach has proved effective for the transmission of contents, SciFinder is still a skill and a procedure by nature, which requires the user to experience it hands-on to master the technique. Therefore it is inevitable that no approach, structure, or curriculum can sufficiently prepare students to use SciFinder at a sufficient level without students' own practice. In addition, it requires the students' cooperation to come to class prepared. Although email reminders were sent to the students, some students might forget to watch the videos and prepare for the F2F session, owing to busy schedules and commitments. We suggest implementing a presession online quiz carrying 1%–2% of the overall grade, which will help to encourage students in watching the videos.

Conclusion

In conclusion, this study presents a method of teaching and learning the use of SciFinder using the flipped classroom approach. With this approach, we saved 25% of time originally for the physical F2F sessions, hence gaining more time for hands-on exercises. This fresh approach has shown considerable improvement in students' grades, with most students agreeing that this approach helped them learn better. Furthermore, this is also easily transferrable to other educators who may wish to adapt this method of teaching how to use SciFinder and other database tools.

Acknowledgement and Declaration

The authors declare that this chapter is not sponsored by SciFinder nor CAS. The authors thank a senior science librarian from NUS Libraries, an expert from ACS, and a former undergraduate student who supported with the editorial process.

References

[1] Love BE, Bennett LJ. Determining synthetic routes to consumer product ingredients through the use of electronic resources. J Chem Educ 2016;93:567–8. https://doi.org/10.1021/acs.jchemed.5b00391.

[2] Ridley D. Introduction to searching with SciFinder Scholar. J Chem Educ 2001;78:557–8. https://doi.org/10.1021/ed078p557.

[3] O'Reilly SA, Wilson AM, Howes B. Utilization of SciFinder Scholar at an undergraduate institution. J Chem Educ 2002;79:524–6. https://doi.org/10.1021/ed079p524.

[4] Somerville AN. SciFinder Scholar (by chemical abstracts service). J Chem Educ 1998;75:959. https://doi.org/10.1021/ed075p959.

[5] Fung FM. Adopting lightboard for a chemistry flipped classroom to improve technology-enhanced videos for better learner engagement. J Chem Educ 2017;94:956–9. https://doi.org/10.1021/acs.jchemed.7b00004.

[6] Weaver MG, Samoshin AV, Lewis RB, Gainer MJ. Developing students' critical thinking, problem solving, and analysis skills in an inquiry-based synthetic organic laboratory course. J Chem Educ 2016;93:847–51. https://doi.org/10.1021/acs.jchemed.5b00678.

[7] Baykoucheva S, Houck JD, White N. Integration of endnote online in information literacy instruction designed for small and large chemistry courses. J Chem Educ 2016;93:470–6. https://doi.org/10.1021/acs.jchemed.5b00515.

[8] Baysinger G. Introducing the Journal of Chemical Education's "special Issue: Chemical Information". J Chem Educ 2016;93:401–5. https://doi.org/10.1021/acs.jchemed.6b00113.

[9] Christiansen MA. Inverted teaching: applying a new pedagogy to a university organic chemistry class. J Chem Educ 2014;91:1845–50. https://doi.org/10.1021/ed400530z.

[10] Jacobs DL, Dalal HA, Dawson PH. Integrating chemical information instruction into the chemistry curriculum on borrowed time: the multiyear development and evolution of a virtual instructional tutorial. J Chem Educ 2016;93:452–63. https://doi.org/10.1021/acs.jchemed.5b00427.

[11] Fung FM, Watts SF. Bridges to the future: toward future ready graduates in chemistry laboratories. J Chem Educ 2019. https://doi.org/10.1021/acs.jchemed.8b00771.

[12] Fung FM, Chng WH, Tan HR, Ng MTT. Sustaining active learning in virtual classroom; 2020. p. 6–7.

[13] Ardisara A, Fung FM. Integrating 360° videos in an undergraduate chemistry laboratory course. J Chem Educ 2018. https://doi.org/10.1021/acs.jchemed.8b00143.

[14] Ashraf SS, Marzouk SAM, Shehadi IA, Brian MM. An integrated professional and transferable skills course for undergraduate chemistry students. J Chem Educ 2011;88:44–8. https://doi.org/10.1021/ed100275y.

[15] Buntrock RE. Using citation indexes, citation searching, and bibliometrics to improve chemistry scholarship, research, and administration. J Chem Educ 2016;93:560–6. https://doi.org/10.1021/acs.jchemed.5b00451.

[16] Currano JN. Introducing graduate students to the chemical information landscape: the ongoing evolution of a graduate-level chemical information course. J Chem Educ 2016;93:488–95. https://doi.org/10.1021/acs.jchemed.5b00594.

[17] Danowitz AM, Brown RC, Jones CD, Diegelman-Parente A, Taylor CE. A combination course and lab-based approach to teaching research skills to undergraduates. J Chem Educ 2016;93:434–8. https://doi.org/10.1021/acs.jchemed.5b00390.

[18] Gozzi C, Arnoux MJ, Breuzard J, Marchal C, Nikitine C, Renaudat A, et al. Progressively fostering students' chemical information skills in a three-year chemical engineering program in France. J Chem Educ 2016;93:576–9. https://doi.org/10.1021/acs.jchemed.5b00414.

[19] Graham KJ, Schaller CP, Jones TN. An exercise to coach students on literature searching. J Chem Educ 2015;92:124–6. https://doi.org/10.1021/ed500486p.

[20] Greco GE. Chemical information literacy at a liberal arts college. J Chem Educ 2016;93:429–33. https://doi.org/10.1021/acs.jchemed.5b00422.

[21] Härtinger S, Clarke N. Using patent classification to discover chemical information in a free patent database: challenges and opportunities. J Chem Educ 2016;93:534–41. https://doi.org/10.1021/acs.jchemed.5b00740.

[22] Rosenstein IJ. A literature exercise using SciFinder Scholar for the sophomore-level organic chemistry course. J Chem Educ 2005;82:652. https://doi.org/10.1021/ed082p652.

[23] Scalfani VF, Frantom PA, Woski SA. Replacing the traditional graduate chemistry literature seminar with a chemical research literacy course. J Chem Educ 2016;93:482–7. https://doi.org/10.1021/acs.jchemed.5b00512.

[24] Shultz GV, Li Y. Student development of information literacy skills during problem-based organic chemistry laboratory experiments. J Chem Educ 2016;93:413–22. https://doi.org/10.1021/acs.jchemed.5b00523.

[25] Tan HR, Chng WH, Chonardo C, Tao M, Ng MTT, Fung FM. How chemists achieve active learning online during the COVID-19 pandemic: using the Community of Inquiry (CoI) framework to support remote teaching; 2020. https://doi.org/10.1021/acs.jchemed.0c00541.

[26] Anuar H, Fung FM. YouTube video titled SciFinder Tutorial: the most comprehensive chemical literature database; 2017.

[27] Anuar H, Fung FM. YouTube video titled SciFinder Registration Tutorial for NUS students; 2017.

[28] Swoger BJM, Helms E. An organic chemistry exercise in information literacy using SciFinder. J Chem Educ 2015;92:668–71. https://doi.org/10.1021/ed500581e.

[29] Yeagley AA, Porter SEG, Rhoten MC, Topham BJ. The stepping stone approach to teaching chemical information skills. J Chem Educ 2016;93:423–8. https://doi.org/10.1021/acs.jchemed.5b00389.

[30] Zwicky DA, Hands MD. The effect of peer review on information literacy outcomes in a chemical literature course. J Chem Educ 2016;93:477–81. https://doi.org/10.1021/acs.jchemed.5b00413.

Using an NMR software as an instructional tool in elucidating organic structures

Max J.H. Tan[a,b], Kevin Christopher Boellaard[b], Shaphyna Nacqiar Kader[b], and Fun Man Fung[b,c]

[a]*Special Programme in Science, National University of Singapore, Singapore, Singapore*
[b]*Department of Chemistry, National University of Singapore, Singapore, Singapore*
[c]*Institute for Applied Learning Sciences and Educational Technology (ALSET), NUS, Singapore, Singapore*

Introduction

Synthesis laboratory classes are ubiquitous in undergraduate chemical education, taking up 20 of 45 laboratory classes at the undergraduate level at the National University of Singapore (NUS) [1]. Such synthesis experiments often have nuclear magnetic resonance (NMR) as a staple characterization method that allows students to contextualize theory by trying their hand at solving realistic spectra [2–5]. Ideally, students should perform NMR analysis and process their individual raw data independently, allowing them not only hands-on experience at handling research-level equipment but also training them to contextualize and make sense of their data independently. However, the costs of NMR units are exceedingly prohibitive, with most universities having limited sets [4]. As a result, samples produced by students are collected after each synthesis class and analyzed en-masse by a trained laboratory technician. Because the output files (e.g., .fid) by the NMR unit cannot be opened without a specialized software, the laboratory technician need to process the data (i.e., peak clean-up, baseline correction, and integration of the relevant peaks) before converting to a general-use-friendly file (e.g., .pdf file), taking days to complete. Although students will eventually receive personalized .pdf files specific to their product and are still able to verify their product's structures, they will not personally interact with their raw data to isolate or eliminate key peaks relevant to an imperfect NMR spectrum. Consequently, owing to the lack of exposure, students may be unaware or inexperienced in handling and parsing realistic NMR data, a skill integral in research. This technology report, thus, aims to highlight a means of exposing students to handling raw NMR data through the use of a software, MestReNova (MNova). By distributing and allowing students' access to MNova, students will be able to gain first-hand experience at handling NMR data while simultaneously reducing logistical and operational costs of NMR analysis for laboratory classes [4].

Background

Characterization of organic synthesis products often requires several forms of spectroscopic analytical methods, with NMR characterization being one of the most common. Consequently, much effort has been made in trying to cultivate NMR-characterization skills by either introducing NMR at the high school level [1] or making NMR education more innovative. Of late, NMR education has taken a turn toward the high-tech, with individuals such as Vosegaard pushing for the use of iSpec [6], a web-based spectroscopy activity, and groups such as Masania et al. introducing detailed, iterative methods to integrate hands-on NMR education [7]. Notably, groups such as Topczewski et al. have even employed eye-tracking technologies to better understand eye movements of students when reading NMR spectra with the hopes of improving NMR education [8]. Despite the laudable efforts by the myriad groups, realism in NMR education remains sparse; most NMR education involved the use of either theoretical or sterile NMR spectra, which are unrepresentative of realistic experiments. Our group, thus, aims to fill this gap by exposing undergraduate students to realistic NMR results to train critical reasoning skills [1, 9]. We therein describe an activity employing and NMR software tool in laboratory classes targeted toward addressing this gap.

Mestrenova

Raw data processing

Distribution

MNova can be easily downloaded and installed from the developer's site [10] without a physical copy of the disk but requires a license file for activation. Notwithstanding, academic institutes can arrange for a distribution license [11] with site licenses directly from the vendor.

Opening files

The crux of this technology report tackles the use of MNova for processing of the raw data obtained from NMR analysis. Using data obtained from Bruker 500 UltraShield as in NMR analysis at NUS, we will outline key steps pertaining to how the raw data can be processed.

On completion of NMR analysis by the laboratory technician, students will be provided personalized sets of data files. Of the collection of files, the user (i.e., the student) can open the "fid" file using MNova. This should open the software and display the NMR spectrum, which can be analyzed using tools available in the top toolbar (outlined). This report will outline functions that are applicable for use by undergraduate students in laboratory class.

Peak isolation

By default, the spectrum displayed will range the intensities and parts per million (ppm) as set by the NMR unit (Fig. 7.1). The user can vary the displayed intensities by scrolling or zoom into specific portions of the spectrum using the zoom function (Fig. 7.2). The ppm of specific peaks can be displayed using several functions. Although the default "peak pick" function displays all the peaks present automatically, peaks are conventionally picked manually or by threshold intensity to isolate experimentally relevant features. This allows the user to pinpoint specific peaks or select a collection of peaks by specifying a range of intensities, respectively. The user will, thus, need to reason and decide as to which peaks should be ignored (e.g., impurities, noise, or solvents). The isolated, relevant peaks can be further analyzed for their coupling constants using the "multiplet analysis" function. This function automatically calculates coupling constants for the isolated peaks.

Peak integration

The "peak integration" function elucidates the relative intensities of selected regions (Fig. 7.2). Likewise, whereas the "peak integration" function automatically integrates selected peaks, regions are conventionally manually isolated because the automatic function is unable to discern multiplets from a collection of peaks. Notwithstanding, the first isolated region will have an integration value set to unity, and every other region thereafter will have integration values relative to the first region. Although principally, the integration ratios should ultimately yield the same results regardless which peak is set as the reference, the ease of elucidation is greatly increased with the correct reference (i.e., peak indicating the smallest number of H atoms).

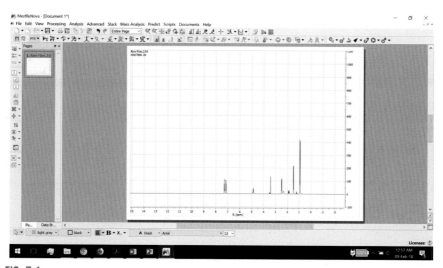

FIG. 7.1

Default page when a ".fid" file is opened using MNova.

FIG. 7.2

Close-up of top toolbar with zoom features outlined in *red* and peak picking features in *yellow*. From left to right, *red*: zoom in, zoom out, full spectrum manual zoom, pan, and expansion. *Yellow*: peak pick, peak integration, and multiplet analysis.

Information presentation

On completion of the data processing, specific regions can be isolated and expanded for emphasis and presentation. This can be done by selecting the "expansion" button and highlighting the region, of which result a pop-up image of the selected region. This image can be maneuvered to empty regions of the spectrum simply by dragging. The user can further construct the chemical structure of the molecule by selecting the relevant tool in the "draw" section. Alternatively, molecules drawn on ChemDraw can be imported onto the spectrum simply by "copying" and "pasting" the molecule onto the spectrum (Fig. 7.3) [12, 13].

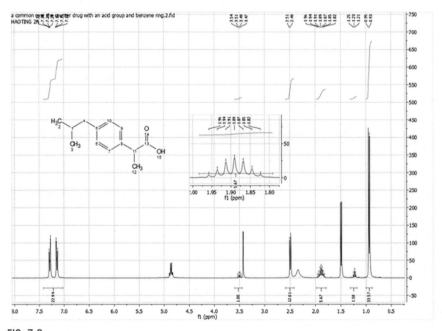

FIG. 7.3

Analyzed spectrum with relevant peaks, integration ratios, expanded section, and structural representation.

Propagation of technology

Besides providing textual instructions for the installation and use of MNova, walk-through videos have been made to facilitate use in the classroom [14] (Fig. 7.4). In addition, the use of MNova synchronizes well with software such as ChemDraw Ultra, which features molecule drawing and spectrum prediction [15].

Discussion

By exposing students to raw experimental data at the undergraduate level, students will be able to attain first-hand experience at handling data imperfections pertinent in research. Examples of such data imperfections relevant to NMR would be identifying solvent peaks, isolating irrelevant, impurity peaks and noise. Furthermore, students can also pick up optimization skills not necessarily taught in the classroom but gained through repetitive practice. As an example, by comparing two spectra of the same compound but from separate runs of the experiment (Fig. 7.5), one can observe similarity and differences in the spectra. The differences in spectra indicate substances that are not present in both runs and could be due to impurities (i.e., these impurity peaks are irrelevant). Similarly, common peaks, although might be representative of the intended compound, could also be solvent peaks. Measured assumptions and guesswork allow students to conduct cost-benefit analyses when ruling in or out certain peaks, thereby honing higher-order reasoning abilities. Finally, peak integration relative to the weakest, relevant peak can optimize the process of deducing how many hydrogen atoms are represented in each peak. Ultimately, exposing students to these imperfections (as opposed to sterile NMR spectra commonly used in the classroom) can potentially train higher-order critical reasoning skills (Fig. 7.5).

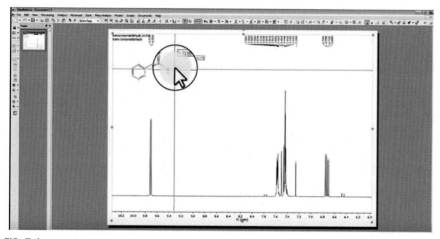

FIG. 7.4

Screenshot of video teaching the basics of how to use MNova.

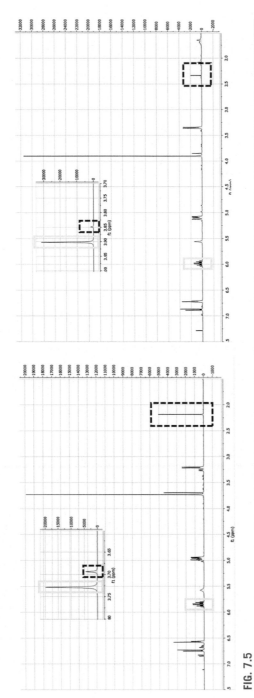

FIG. 7.5

Sample spectra of the same product synthesized separately by different experimenters. Common peaks are outlined in *yellow*, insinuating that they are peaks of products common in both samples. Uncommon peaks are outlined in *dashed red boxes*, suggesting that they could be from impurity analytes.

Conclusions

The use of technology to bring students closer to experimental data can give them a head-start in research. We have outlined a method for the use of MNova by undergraduate students to handle raw NMR data. This method can aid students in developing higher-order critical reasoning skills by requiring them to conduct cost-benefit analyses or measured assumptions in solving imperfect data. This early exposure also trains students in developing optimization skills not necessarily taught in the classroom.

Acknowledgments

The authors thank #ChemNUS for acquiring MNova in CM3291. We appreciate Hoang Truong Giang, Han Yan Hui, and Wu Ji'en for their effort in the NMR initiatives. Thanks also to Ong Jun Yang for his contribution in the MNova installation manual. The authors declare no sponsorships and incentives from any of the programs referenced within the article.

References

[1] Jacob C. Critical thinking in the chemistry classroom and beyond. J Chem Educ 2004;81(8):1216–23.

[2] Hilty C, Bowen S. An NMR experiment based on off-the-shelf digital data-acquisition equipment. J Chem Educ 2010;87(7):747–9.

[3] Chamberlain PH. Identification of an alcohol with ^{13}C NMR spectroscopy. J Chem Educ 2013;90(10):1365–7.

[4] Mills NS, Shanklin M. Access to NMR spectroscopy for two-year college students: the NMR site at Trinity University. J Chem Educ 2011;88(6):835–9.

[5] Iler HD, Justice D, Brauer S, Landis A. Discovering ^{13}C NMR, ^1H NMR, and IR spectroscopy in the general chemistry laboratory through a sequence of guided-inquiry exercises. J Chem Educ 2012;89:1178–82.

[6] Vosegaard T. iSpec: a web-based activity for spectroscopy teaching. J Chem Educ 2017. https://doi.org/10.1021/acs.jchemed.7b00482.

[7] Masania J, Grootveld M, Wilson PB. Teaching analytical chemistry to pharmacy students: a combined, iterative approach. J Chem Educ 2018;95(1):47–54.

[8] Topczewski JJ, Topczewski AM, Tang H, Kendhammer LK, Pienta NJ. NMR spectra through the eyes of a student: eye tracking applied to NMR items. J Chem Educ 2017;94(1):29–37.

[9] Stowe RL, Cooper MM. Practicing what we preach: assessing "critical thinking" in organic chemistry. J Chem Educ 2017;94(12):1852–9.

[10] Research M. http://mestrelab.com/download/mnova/ http://mestrelab.com/download/mnova/ [Accessed February 3, 2018].

[11] National University of Singapore Chemistry Department. Installation MNOVA. https://chemistry.nus.edu.sg/cmmac/nmr/MestReNova%20Installation%20Guide%20F.pdf [Accessed February 5, 2018].

[12] Morsch LA, Lewis M. Engaging organic chemistry students using ChemDraw for iPad. J Chem Educ 2015;92(8):1402–5.

[13] Li Q, Chen Z, Yan Z, Wang C, Chen Z. Touch NMR: an NMR data processing application for the iPad. J Chem Educ 2014;91(11):2002–4.

[14] Fung FM. MestreNova tutorial: a quick and effective guide on NMR processing, https://www.youtube.com/watch?v=BuePpQWgEVw. [Accessed February 26, 2018].

[15] PerkinElmer. ChemDraw and ChemOffice. https://www.cambridgesoft.com/software/overview.aspx [Accessed March 12, 2018].

Case studies

Flipped chemistry in multisite IVC courses: A possible model for the future of virtual chemistry education

Michael A. Christiansen

Department of Chemistry and Biochemistry, Utah State University – Uintah Basin Campus,
Vernal, UT, United States

Interactive videoconferencing courses might serve as a model for classes attended over computers or smartphones through videoconferencing apps

Recent global events have rapidly made online course delivery a sweeping reality for the foreseeable future, with educators at every institutional level being thrust into teaching classes through videoconferencing apps such as Zoom, Cisco Webex, Adobe Connect, and others. In the face of such a fundamental upheaval, educators who are accustomed to a traditional lecture (TL) format might feel understandably reticent about also flipping or blending such courses because this would add one *more* layer of complexity to an already difficult task. However, for those who are considering such an undertaking, as well as any who have flipped previously and wonder how to adapt it to their present circumstance, I offer lessons gained from 6 years of flipping college chemistry classes that were delivered in live, multisite, interactive videoconference (IVC) settings.

With 32 campus or learning centers spread across the approximately 85,000-square-mile state of Utah [1], Utah State University (USU) has offered electronically delivered distance courses and programs since 1984, making it an institutional world leader in distance education. IVC technology was originally incorporated into USU's distance education system in the early 2000s [2].

Although my past work in this area [3–6] is not *identical* to flipping a pure videoconference class given over computer or smartphone, it does possess parallels that might serve as a reasonable comparative model. After all, in multisite IVC courses, instructors live-teach geographically dispersed students who are visible on a split-screen display, similar to a videoconference environment. Although *flipping* such

Technology-Enabled Blended Learning Experiences for Chemistry Education and Outreach.
https://doi.org/10.1016/B978-0-12-822879-1.00008-1

117

courses might seem unnecessarily complex, I have achieved positive results while doing so over the past 6 years [3–6]. Thus to help any who intend to flip future videoconferencing classes to avoid pitfalls, this chapter includes the results of extensive data analysis drawn from my past IVC flipping experience.

What is flipped learning? And what are multisite, synchronously delivered, geographically dispersed courses?

The terms "flipped learning" [7] or "flipped classroom" [8]—known also as "classroom flip" [9, 10], "backwards classroom" [11], "reverse teaching" [12], "upside-down classroom" [13], "flip teaching" [12], "inverted classroom" [14–16], and "inverted teaching" [3]—reference a pedagogy in which students watch prerecorded lectures outside class and spend in-class time carrying out instructor-facilitated problem sets or other higher-learning activities [17–21]. Thus the flipped classroom swaps the *locations* of lecture and homework.

This method was originally proposed in 2000 by Baker [22] and separately by Lage, Platt, and Treglia [14, 15]. With a few noteworthy exceptions that include Refs. [9, 10, 13], flipping did not gain much attention in peer-reviewed literature until 2011 [18, 23], with the number of articles rising steadily ever since [24].

Although some practitioners may disagree, I personally define a *flipped course* as one that maintains the same number of class periods as its TL counterpart, with only the *locations* of its lecture and homework (or other activities) trading places. This contrasts with a *blended* course, whose total number of class periods is *lower* than its TL counterpart because some classes get replaced by fully online activities. Thus *blended courses* are a hybrid of fully online learning and traditional in-class meetings or lectures, thereby differing somewhat from a fully flipped class.

Despite flipped learning's growing literature exposure, its application to multisite, synchronously delivered, geographically dispersed courses has received limited coverage. As I foreshadowed earlier, to help increase rural students' access to higher education, many universities have built geographically dispersed satellite campuses that are networked through broadcast IVC technology [1, 25–34]. This allows instructors to teach students simultaneously across multiple locations—sometimes hundreds of miles apart—thereby enabling place-bound individuals in rural areas to earn previously inaccessible degrees [35]. Students can attend such courses in-person (called *face-to-face* or "F2F" students) or from geographically distant sites (called "distance students"). Although F2F students have a more traditional in-class experience, distance students see their teacher from diverse locations in real time on a projected screen. The instructor, in turn, sees *students* on a split screen display at the back of the room. All verbal communications with distance students are conducted through microphone [36].

This format is pedagogically challenging because the instructor can only *personally* observe *F2F* students, but not *distance* students, with whom interactions are limited to electronic communication through screens, microphones, virtual chat rooms, or emails. According to *transactional distance theory* [6, 37–39], this literal geographic distance may exacerbate feelings of detachment between the instructor and distance students and thereby contribute to disengagement and decreased learning. This concern notwithstanding, I have been delivering flipped IVC courses since fall 2013, having published two peer-reviewed articles and one book chapter on the subject [4–6]. Outside of *my* work, however, I have encountered just one other peer-reviewed article that specifically addresses this unique topic of flipping a university IVC course [38].

Literature criticisms of flipped learning are limited but merit addressing

At the 2016 ACS National Meeting, I gave a symposium talk and poster presentation on flipped learning [40, 41]. In one fascinating exchange, a professor looked at my poster for a few minutes and then curtly said, "I despise flipped learning." Unflustered, emotionless, but genuinely intrigued, I asked her to explain. She responded, "It's a fleeting trend that doesn't accomplish anything new."

Sadly, I was unable to get more information before she left. Nevertheless, a thorough reading of many peer-reviewed articles on the subject shows a general literature trend of speaking only *complimentarily* about flipping, while addressing the pedagogy's shortcomings or failures just briefly, if at all. In fact, one article asserts that "negative academic results from flipped classrooms are rarely published" [23].

Accurate or not, the fact remains that to improve *any* method, one cannot simply dismiss or ignore its shortcomings and failures but must tackle them head-on. Thus to help increase awareness of the drawbacks or deficiencies of flipping, for the purpose of better addressing them, my thorough (albeit likely not comprehensive) scouring of the literature revealed several consistent themes.

Faculty concerns

Faculty who resist flipping their courses do so for various reasons, which include concerns about (a) the time investment it will require [15, 38, 42, 43], (b) what to do in class [44], (c) worrying that students might learn fewer fundamentals than in a TL structure [45], (d) using a technocentric fad with a short track record [44], and (e) not having experienced flipped learning as students themselves [45]. Rotellar and Cain also postulate that "some faculty … may struggle with releasing the reliance on their role as content-deliverer, especially if they have been considered by themselves or others to be a 'great teacher'" [45]. Furthermore, traditional lecture halls are often poorly designed for flipped courses in which students need to sit in circles for group work [23].

One thorough study [46], whose findings were corroborated by two later reports [19, 47], indicated that the grade and attitude improvements that result from a flipped format arise from the *active learning* component of flipping, not the flipping itself [46]. Thus, incorporating *active learning* into a traditional lecture (TL) course may yield similar gains [46]. In addition, although numerous reports suggest that flipping may improve academic performance and attitudes relative to TL [3, 17, 19, 47–49], there are some exceptions [38, 50].

Student concerns

Although a well-designed flipped course does not require more overall time of students than TL [23], uninitiated students often *worry* that it will, which can cause anxiety—especially among nontraditional students with less time availability outside of class [23]. However, numerous findings indicate that once they have experienced both, most students prefer flipping to TL [15, 51–54]. The following additional concerns also arise repeatedly in the literature:

1. Students who are new to flipping sometimes resist it for a variety reasons, which include presumptions that flipping will (a) shift the burden of learning more to students [55], (b) require them to do more work [56], (c) "equate to students teaching themselves while faculty members stand idly by" [45, 57], or (d) result in "a chaotic classroom environment" [10], or "classroom unsettledness" [45]. In addition, new students may worry that their previous study habits of memorizing material, cramming for examinations, and forgetting it afterward will not succeed in a flipped course [45].
2. When at-home videos are too long, students may dislike flipped courses and thereby forfeit potential learning gains [58]. The literature consensus seems to place the ideal average video length at around 10 min or less [17, 48, 59].

To help determine how well my student feedback aligns with or deviates from the above-listed concerns, I thoroughly analyzed 8 years of anonymous survey data. None of the abovementioned "faculty concerns" were taken into account in my work. I instead focused solely on student concerns. Overall, my data *did* support the items listed as 1 and 2 just above and also proffered additional considerations. Although some of these are specific to my course design, teaching approach, and management style, I recommend contemplating them for *any* flipped course, especially one in an IVC or virtual setting.

Methods

Number of courses flipped

As Fig. 8.1 indicates further, I teach four university courses through a flipped classroom structure: freshman General Chemistry I and II (abbreviated as GC I and GC II) and sophomore Organic Chemistry I and II (abbreviated as OC I and OC II). All these are four-credit courses that were delivered in multisite IVC settings to a mixture of F2F and distance students. I teach GC I and II annually (Fig. 8.1, Items 1–2) but did not flip them until fall 2014 [6]. By contrast, I only teach OC I and II biennially (Fig. 8.1, Items 3–4), with my first OC I being conducted through TL in fall 2011

Item	Course (abbreviation)	Delivery frequency (term)	Total semesters flipped
1	General Chemistry I (GC I)	Annual (fall)	6 (fall 2014–fall 2019)
2	General Chemistry II (GC II)	Annual (spring)	6 (spring 2015–spring 2020)
3	Organic Chemistry I (OC I)	Biennial (fall of odd years)	3 (fall 2013–fall 2017)[a,b]
4	Organic Chemistry II (OC II)	Biennial (spring even years)	4 (spring 2012–spring 2018)[b]

[a] OC I in fall 2011 is not included because it was delivered via TL.

[b] Although flipped, OC I in fall 2019 and OC II in spring 2020 are not included because a new and significantly different digital textbook was used.

FIG. 8.1

Flipped courses taught.

(Fig. 8.1, footnote *a*) and every OC course thereafter being flipped [3]. Moreover, although OC I and OC II in fall 2019 and spring 2020 were flipped (Fig. 8.1, footnote *b*), we will not consider them here because they involved a new digital textbook that differed significantly from previous iterations, thereby making them too dissimilar for direct comparison.

Course designs and structures

As we foreshadowed above and described thoroughly in previous work [5, 6], each course in Fig. 8.1 was delivered to a mixture of F2F and distance students. Although F2F students attended in-person with the instructor, distance students viewed class live from various satellite campuses across Utah, connected through the public *Utah Education Network* [60]. This allowed distance students to see and hear the instructor, lecture slides, and overhead ELMO Cam notes in real time on a projected screen at the front of their rooms. Distance sites were further equipped with microphones that enabled students to ask questions live and participate in real-time discussions. At the origination site, the instructor could see distance students on a split screen display at the back of the room and toggle between video capture of lecture slides or ELMO Cam notes, as needed.

Each course in Fig. 8.1 spanned 14 weeks and included two 100-min weekly class periods that covered about one textbook chapter per week (see the Supporting Information of Refs. [3, 5, 6] to access textbook, syllabi, and question lists). Students were required to watch lecture videos before class and individually take weekly in-class quizzes, according to schedules posted on the syllabi. Students then worked in assigned groups of two to five people on mixed-format (open-answer,

Course	No. of videos (average length)	In-class activities[a] (% of course grade)	Exams[a,b] (% of course grade)
GC I	99 (7.5 min)	12 In-class quizzes (~9%) 12 Problem sets (~18%)	Entrance exam (~6%) 4 Midterms (50%) Comprehensive final (~17%)
GC II	143 (6.5 min)	12 In-class quizzes (~10%) 12 Problem sets (~19%)	Intro survey (~1%) 4 Midterms (52%) Comprehensive final (~18%)
OC I	79 (9.5 min)	13 In-class quizzes (10%) 13 Problem sets (20%)	Survey and entrance exam (~4%) 4 Midterms (50%) Comprehensive final (~16%)
OC II	65 (8.8 min)	13 In-class quizzes (~10%) 13 Problem sets (~21%)	Intro survey (~1%) 4 Midterms (~51%) Comprehensive final (~17%)

[a] For each class, the student's lowest problem set, lowest quiz, and lowest midterm scores were dropped.

[b] Exams were taken at proctored university testing centers.

FIG. 8.2

Each course's video number, average length, and grade percent for all quizzes, problem sets, and examinations.

multiple-choice, etc.) problem sets that were written by the instructor or adapted from published question banks. Fig. 8.2 shows the total number of videos, average video lengths, and number of quizzes, problem sets, examinations, and other activities for each course, with their respective grade percent breakdowns. As footnote *a* of Fig. 8.2 indicates, students' lowest problem set, quiz, and midterm scores were dropped at the end of each term. Moreover, although quizzes and problem sets were performed in class, examinations were taken at proctored university testing centers (see Fig. 8.2, footnote *b*), which freed up time for group work and discussions.

Each group member received the same problem set grade. However, individual problem set grades were adjusted at the end of the term to incorporate students' anonymous "peer grades," as described in previous work [3, 6]. Videos were made using varied versions of Camtasia Studio and included PowerPoint lectures with picture-in-picture footage of the narrating instructor or of the instructor solving problems on a board. During class, the instructor intermittently observed student work. For distance students, this was conducted by checking in over the microphone every 5–7 min. As questions arose, the instructor either directed students to helpful resources or gave mini-lectures to clarify confusion. Throughout the semester, students were allowed to submit problem sets at any time until the due date and get feedback without losing points. Distance students did this electronically (through email) as scanned PDFs or photographs taken with smartphones. The instructor then gave feedback straightway over the ELMO Cam or made annotations to student submissions on his computer and replied to the entire distance group immediately through email. This feedback question-answer loop encompassed most of class time.

Population studied

Over the span of the 6 years and the various courses listed in Fig. 8.1, our treatment (class delivery) involved a total of 120 (47 + 32 + 21 + 20) F2F students and 358 (144 + 101 + 49 + 64) distance students, as Fig. 8.3 indicates. The small number of F2F Chemistry majors (Fig. 8.3, middle row) results primarily from the unavailability of the full Chemistry bachelor's degree at our isolated satellite campus. However, even among my *distance* students from the central campus, very few Chemistry majors have taken my courses (Fig. 8.3, middle row), underscoring the fact that these are primarily service courses given to nonmajors. In addition, for simplification, self-reported student ethnicities were amalgamated and averaged across *all* four courses (GC I, GC II, OC I, and OC III), as seen in the middle section of Fig. 8.3.

Survey instruments

As detailed in previous work [3], student opinions were formatively assessed through two anonymous survey instruments per semester: one at week 4 (soon after examination 1)

Item (SD = standard dev.)	GC I (F2F, distance)	GC II (F2F, distance)	OC I (F2F, distance)	OC II (F2F, distance)
Total courses analyzed	6	6	3	4
N (total: 120 F2F, 358 distance)	47, 144	32, 101	21, 49	20, 64
Gender (M = male, F = female)	26 M, 93 M 21 F, 51 F	15 M, 57 M 17 F, 44 F	16 M, 33 M 5 F, 16 F	15 M, 46 M 5 F, 18 F
Mean age, years (SD)	26 (6.6), 26 (6.1)	29 (7.8), 24 (5.1)	25 (7.2), 24 (4.7)	29 (7.4), 24 (4.8)
Chemistry majors[a] (%)	0.0, 7.6	9.4, 7.9	0.0, 8.2	5.0, 14.1
Ethnicity[b]	White/Caucasian Hispanic or Latino Asian Native American or Alaska Native Black or African American Other or unspecified		F2F: 88.6%, distance: 89.1% F2F: 4.19%, distance: 4.38% F2F: 1.40%, distance: 1.25% F2F: 1.40%, distance: 0.94% F2F: 1.12%, distance: 1.25% F2F: 3.35%, distance: 3.13%	
Mean GPA (SD)	3.26 (0.46), 3.24 (0.63)	3.40 (0.43), 3.28 (0.51)	3.44 (0.38), 3.50 (0.31)	3.45 (0.33), 3.50 (0.31)
Median GPA	3.28, 3.32	3.48, 3.27	3.45, 3.60	3.30, 3.30

[a] The full chemistry degree program is unavailable at our F2F location.

[b] For simplification, ethnicity percentages have been combined for all courses.

FIG. 8.3

Student demographic data.

and another at week 14. Each evaluation, administered through Canvas (our course management system or CMS), had quantitative (Likert scale) and qualitative (open-ended) sections. Although voluntary, participation was incentivized by five extra-credit points, thereby resulting in a >95% participation rate. The CMS *did* allow the instructor to distinguish between F2F and distance student feedback; all results were otherwise kept completely deidentified/anonymous.

Results and discussion
Quantitative findings

Two surveys were given per semester for each class, and each survey posed 14–16 questions per student. Thus despite having a total of only 120 F2F students and 358 distance students over 6 years (Fig. 8.3), the surveys still generated thousands of individual data point responses, as shown in Fig. 8.4.

Careless in-depth analysis of such a large data set could produce results that are complex, meandering, or unclear. To avoid this, Fig. 8.5 summarizes responses to the four key survey questions that are of greatest value to our present objectives, with averages amassed across all 19 of the courses originally shown in Fig. 8.1 (six GC I + six GC II + three OC I + four OC II). It should be noted:

- Surveyed students had never experienced TL with the same instructor, so they were only comparing their experience with flipped learning to past TL general experiences with *different* instructors (Fig. 8.5, Item 1).
- Results for the final question (Fig. 8.5, Item 4) only include responses from students who had experienced an IVC course previously. Because USU is a leading innovator in distance education, 70% or more of USU students have experienced an IVC course before.

As Fig. 8.5 clearly shows, a strong majority of students recorded a 4 or 5 ("somewhat agree" or "strongly agree") in response to these four key questions, which

	Week 4	Week 14
[a]Quantitative (Likert-scale responses)	3846	3474
[a]Qualitative (open-ended questions)	2814	3035

[a]F2F and distance numbers combined.

FIG. 8.4

Total numbers of survey questions answered.

Item (TL = traditional lecture)		Strongly disagree, % (F2F, distance)	Somewhat disagree, % (F2F, distance)	Neither agree nor disagree, % (F2F, distance)	Somewhat agree, % (F2F, distance)	Strongly agree, % (F2F, distance)
I prefer flipped learning to a TL format[a]	Week 4:	3.3, 3.2	0.0, 3.9	13.3, 14.1	13.3, 40.2	70.0, 38.6
	Week 14:	0.0, 5.0	0.0, 4.5	2.4, 12.6	18.1, 28.1	79.6, 49.9
I like working in groups	Week 4:	2.5, 2.9	3.8, 9.3	16.5, 13.7	38.8, 38.8	50.0, 45.6
	Week 14:	0.0, 1.5	2.2, 10.8	16.3, 15.2	27.0, 38.6	54.4, 37.9
I like this course	Week 4:	0.0, 3.3	4.2, 5.8	7.5, 18.8	30.0, 41.4	58.3, 30.7
	Week 14:	0.0, 6.4	0.0, 1.8	1.9, 17.1	16.9, 29.6	81.3, 48.4
This course is better than other IVC broadcast courses I've had[b]	Week 4:	0.0, 0.0	0.0, 4.1	8.8, 22.5	13.8, 22.7	77.4, 50.6
	Week 14:	0.0, 0.8	0.0, 5.1	0.0, 9.8	23.3, 16.3	76.7, 68.0

[a] Students surveyed had never experienced TL with the same instructor previously, so they are only comparing their experience with flipped learning to past TL experiences with *different* instructors.
[b] This question only includes results from students who had experienced an IVC course previously.

FIG. 8.5

Quantitative Likert scale survey results, averaged across 19 different courses.

revealed a general trend of: (1) preferring flipped learning to TL (Item 1), (2) liking group work (Item 2), (3) liking the course (Item 3), and (4) finding the course to be better than previous IVC courses they had experienced. These latter data are important because they show that flipping does not intrinsically decrease student perceptions of a course's quality, even in an IVC setting. If this translates to a videoconference course delivered through computer or smartphone, then flipping such courses should not decrease students' perception of course quality or student evaluation scores.

Some responses moved in the more favorable direction between weeks 4 and 14. For example, 3.3% of F2F students strongly disliked flipped learning at week 4 (Item 1, upper left), but that number shrunk to 0.0% by week 14 (Item 1, lower left). This coincides with previous work indicating that students who are new to flipping require an adjusting period to become comfortable with it [11]. However, the *opposite* trend held true for distance students (Item 1, left side).

Qualitative findings

After extensive and thorough analysis of our qualitative data, we found that only 185 (6.57%) of the 2814 responses to week 4 included negative feedback. This percent declined over time, with only 126 (4.15%) of the 3035 responses being negative by week 14. Thus, in general, the flipped course structures we used were positively received. However, as specified earlier, we refuse to simply dismiss or ignore our methods' shortcomings, but instead aim to tackle them head-on. Although proposing solutions for *every* listed concern is beyond this chapter's scope and may be addressed in later work, we did dissect the negative comments, which ended up being codified into 34 types across 10 categories, as shown in Fig. 8.6.

Discussion

The highest-percent items that either remained similar or increased notably between weeks 4 and 14 are boxed in the rightmost columns of Fig. 8.6. The largest of these were complaints about the content being too difficult or being covered too quickly (Fig. 8.6, Item 32). Unfortunately, this may be an intractable obstacle; after all, chemistry is a naturally difficult subject. Nevertheless, complaints about this issue gratifyingly diminished from 15.14% at week 4 to 7.94% at week 14 (Fig. 8.6, Item 32, rightmost columns).

Item 12 arises from perceptions of student-teacher detachment among distance students, which may stem from the literal geographic distance that is incident to IVC education and embodied by *transactional distance theory*, as discussed in previous work [33]. This frequency disappointingly increased from 3.78% at week 4 to 6.35% at week 14 (Fig. 8.6, Item 12, rightmost columns).

Item	Category	Student feedback	Week 4	Week 14
1	Student learning, time management outside of class	Flipping shifts burden of learning to students (concern #1a)	1.08%	0.00%
2		Flipping requires too much work (concern #1b)	0.54%	0.00%
3		Students teach themselves while faculty stand idly by (concern #1c)	1.08%	0.79%
4		If you fall behind, it's hard to catch up	1.08%	2.38%
5		Difficulty finding time outside of class to watch videos	3.24%	4.76%
6		Did not watch videos before class (watched during class)	0.54%	2.38%
7		Difficulty paying attention to videos for extended periods	2.16%	0.00%
8	Problem sets	Want more problems for practice	7.57%	6.35%
9		Want fewer problems	0.54%	0.00%
10	Quizzes	Want quizzes to be online	0.54%	0.79%
11		Want more quizzes	0.54%	0.00%
12	IVC	Students felt distanced from instructor (transactional distance theory)	3.78%	6.35%
13	Difficulties working in groups	Feel there were too many "freeloaders" in their group	1.62%	10.32%
14		Feel their group "divided and conquered" work instead of learning material	1.62%	3.97%
15		Suggests letting students pick their groups	0.00%	0.79%
16		I dislike groups or prefer to work alone	2.16%	5.56%
17		Dislike group work because I'm too far behind or too far ahead of my group	7.03%	7.94%
18	Classroom management	Class time felt like a waste of time or was boring	3.78%	8.73%
19		Classroom management (don't let students monopolize time)	1.08%	1.59%
20		Teacher helped too much instead of just "guiding"	2.16%	0.00%
21		Flipping causes chaotic classrooms (concern #1d)	6.49%	4.76%
22		Want more of class time reviewing content	7.03%	0.79%
23	Videos	The videos were too long (concern #2)	0.00%	1.59%
24		In-class problem sets differ too much from video content	4.86%	1.59%
25		Dislike not being able to ask a video questions	4.86%	7.14%
26		Request that the videos be made in further in advance	2.70%	0.00%
27		Want more problems worked out on the videos	4.32%	0.00%
28		Videos need updating or improved quality	5.95%	7.14%
29		Videos need better organization or functioning links	1.08%	3.97%
30	Exams	Want exams in class	0.54%	0.00%
31		Want more study guides	1.08%	0.00%
32	Content	Content was too hard or covered too quickly	15.14%	7.94%
33	Other	Want an online discussion forum	1.62%	0.00%
34		Unspecified dislike	2.16%	1.59%

FIG. 8.6

Average percent breakdowns of all negative comments, which comprised 6.57% (week 4) and 4.15% (week 14) of all feedbacks.

The higher-frequency responses that lie somewhat beyond the instructor's purview included student concerns about their time management skills (Items 4–6), the aforementioned transactional distance (Item 12), one's inability to ask a video questions (Item 25), and the chaotic classroom environment caused by the flipped structure (Item 21, corresponding with concern #1d). The remaining high-frequency shortcomings (Items 8, 13–14, 16–18, 21, 28–29), however, at least somewhat within the instructor's control, should be taken into account by any interested in flipping their IVC or videoconferencing courses in a similar manner.

Although not of the highest incidence, the most common complaints centered on working in groups (Items 13–14, 16–17). Contrary to one literature suggestion [18], only 0.79% of our student responses suggested letting students pick their own groups. Concerns about groups notwithstanding, my past work showed that when students are allowed to enter or exit groups voluntarily, attendance and academic performance decline [33]. Thus I do not recommend completely abandoning groups. However, providing the option for students to work on their own in extenuating circumstances is acceptable [33].

Conclusions

Our results might attenuate worries among those who are considering flipping a multisite IVC class—or more plausibly—a class delivered in a fully online videoconferencing environment. After all, multisite IVC course delivery is very similar to a videoconferencing virtual class in that IVC instructors teach geographically dispersed students that are visible to the instructor on a split screen display, in real time. Although fears of the potential negative impacts of such bold course restructurizng are understandable, our results indicate that our course design does not decrease the reported course quality, student satisfaction level, or evaluation results.

Nevertheless, even our successes have their shortcomings. Among highly cited concerns over which the instructor has at least some amount of control, qualitative feedback indicated that 1.62%–10.32% of students disliked working in groups for various reasons. These included feeling that there were too many "freeloaders," that their group merely "divided and conquered" instead of learning the material, that individuals were either too far behind or too far ahead of their peers, or that there was a dislike of group work in general. Additional noteworthy complaints centered on the need to periodically update the videos and/or improve their quality or the way they were organized, especially when they included intravideo links that no longer functioned. To the extent that IVC virtual videoconference teaching mirrors the flipped IVC model, as well as the degree to which instructors address these specific issues, their success and student satisfaction rates should exceed ours.

The future of flipped learning is bright and exciting. However, this may just be the beginning. As Enfield explained in 2013: "flipped learning may be one possible step towards a more customized learning environment" [61]. Thus flipping may ultimately serve as a mere transitory bridge toward higher and more deeply impactful forms of tech-centric education, such as virtual reality-driven concept delivery, which may eventually lead to realistic undergraduate research experiences, which are known to increase student retention and success [62–65]. Whatever the case, flipping is and will continue to serve as a valuable conduit toward a brighter educational future, which we now have the special opportunity to help pioneer.

References

[1] Utah State University Regional Campuses and Distance Education Homepage, http://distance.usu.edu/; 2020. [Accessed July 30, 2020].

[2] Barton J. Electronic delivery and technology development. In: Upon the shoulders of giants: a history of Utah State University's Regional Campus System 1967–2015. Logan: Utah State University Provost's Office; 2016. p. 89–93.

[3] Christiansen MA. Inverted teaching: applying a new pedagogy to a university organic chemistry class. J Chem Educ 2014;91:1845–50. https://doi.org/10.1021/ed400530z.

[4] Christiansen MA. Flip teaching college chemistry in broadcast classrooms. In: Blackstock A, Straight N, editors. Interdisciplinary approaches to distance teaching: connecting classrooms in theory and practice. New York: Routledge; 2015. p. 65–86.

[5] Christiansen MA, Lambert AM, Nadelson LS, Dupree KM, Kingsford TA. In-class versus at-home quizzes: which is better? A flipped learning study in a two-site synchronously-broadcast organic chemistry course. J Chem Educ 2017;94:157–63. https://doi.org/10.1021/acs.jchemed.6b00370.

[6] Christiansen MA, Nadelson LS, Cuch MM, Etchberger LH, Kingsford TA. Flipped learning in synchronously-delivered, geographically-dispersed general chemistry classrooms. J Chem Educ 2017;94:662–7. https://doi.org/10.1021/acs.jchemed.6b00763.

[7] Bokosmaty R, Bridgeman A, Muir M. Using a partially flipped learning model to teach first year undergraduate chemistry. J Chem Educ 2019;96:629–39. https://doi.org/10.1021/acs.jchemed.8b00414.

[8] Bergmann J, Sams A. Flip your classroom. International Society for Technology in Education: Eugene, OR; 2012.

[9] Demetry C. Work in progress—an innovation merging "classroom flip" and team-based learning. In: Proceedings of the 40th IEEE frontiers in education conference (FIE), Washington, DC, October 27–30; 2010. TIE1–2.

[10] Strayer JF. The effects of the classroom flip on the learning environment: a comparison of learning activity in a traditional classroom and a flip classroom that used an intelligent tutoring system [Ph.D. thesis]; 2007.

[11] Houston M, Lin L. Humanizing the classroom by flipping the homework versus lecture equation; 2012. p. 1177–82. https://doi.org/10.1-880094-92-4.

[12] Moroney SP. Flipped teaching in a college algebra classroom an action research project, http://scholarspace.manoa.hawaii.edu/handle/10125/27140; 2013. [Accessed July 16, 2020].

[13] Berque D, Byers C, Myers A. Turning the classroom upside down using tablet PCs and DyKnow Ink and audio tools. In: Reed R, Berque D, Prey J, editors. The impact of tablet PCs and pen-based technology on education. Purdue: Purdue University Press; 2009. p. 3–9.

[14] Lage MJ, Platt G. The internet and the inverted classroom. J Econ Educ 2000;31:11.

[15] Lage MJ, Platt GJ, Treglia M. Inverting the classroom: a gateway to creating an inclusive learning environment. J Econ Educ 2000;31:30–43.

[16] Talbert R. Inverted classroom. Colleagues 2012;9:1–2.

[17] Bancroft SF, Fowler SR, Jalaeian M. Leveling the field: flipped instruction as a tool for promoting equity in general chemistry. J Chem Educ 2020;97:36–47.

[18] Casselman MD, Atit K, Henbest G, Guregyan C. Dissecting the flipped classroom: using a randomized controlled trial experiment to determine when student learning occurs. J Chem Educ 2020;97:27–35.

[19] Hibbard L, Sung S, Wells B. Examining the effectiveness of a semi-self-paced flipped learning format in a college general chemistry sequence. J Chem Educ 2015;93:24–30.

[20] Petillion RJ, McNeil WS. Johnstone's triangle as a pedagogical framework for flipped-class instructional videos in introductory chemistry. J Chem Educ 2020;97:1536–42.

[21] Tucker B. The flipped classroom. Educ Next 2012;12:82–3.

[22] Baker JW. The "classroom flip": using web course management tools to become the guide by the side; 2000. p. 9–17.

[23] Jarvis CL. The flip side of flipped classrooms: popular teaching method doesn't always work as planned. Chem Eng News Glob Enterp 2020;98:23–5.

[24] Cheng S-C, Hwang G-J, Lai C-L. Critical research advancements of flipped learning: a review of the top 100 highly cited papers. Interact Learn Environ 2020. https://doi.org/1 0.1080/10494820.2020.1765395.

[25] Bertscha TF, Callasb PW, Rubina A, Caputoc MP, Riccid MA. Applied research: effectiveness of lectures attended via interactive video conferencing versus in-person in preparing third-year internal medicine clerkship students for clinical practice examinations (CPX). Teach Learn Med 2007;19:4–8.

[26] Fonseca JW, Bird CP. Under the radar: branch campuses take off. Univ Bus Mag 2007.

[27] Kondro W. Eleven satellite campuses enter orbit of Canadian medical education. Can Med Assoc J 2006;175:461–2.

[28] Penn State Campuses Homepage; n.d. http://www.psu.edu/academics/campuses [Accessed July 16, 2020].

[29] Texas Tech University System: Campuses & Academic Sites Homepage; n.d. http://www.texastech.edu/campuses.php [Accessed July 16, 2020].

[30] University System of Maryland Homepage; n.d. http://www.usmd.edu/institutions/ [Accessed July 16, 2020].

[31] University of North Carolina, "Our 17 Campuses" Homepage; n.d. https://www.northcarolina.edu/?q=content%2Four-17-campuses [Accessed July 16, 2020].

[32] University of Washington, Bothell Campus Homepage; n.d. http://www.bothell.washington.edu/ [Accessed July 16, 2020].

[33] University of Washington, Seattle Campus Homepage; n.d. http://www.washington.edu/about/ [Accessed July 16, 2020].

[34] University of Washington, Tacoma Campus Homepage; n.d. http://www.tacoma.uw.edu/ [Accessed July 16, 2020].

[35] Hoyt J, Howell S. Why students choose the branch campus of a large university. J Contin High Educ 2012;60:110–6.

[36] Blackstock A, Straight N, editors. Interdisciplinary approaches to distance teaching: connecting classrooms in theory and practice. New York, NY: Routledge; 2015.

[37] Lear JL, Ansorge C, Steckelberg A. Interactivity/community process model for the online education environment. J Online Learn Teach 2010;6:71–7.

[38] McLaughlin JE, Griffin LM, Esserman DA, Davidson CA, Glatt DM, Roth MT, et al. Pharmacy student engagement, performance, and perception in a flipped satellite classroom. Am J Pharm Educ 2013;77. Article 196.

[39] Stein DS, Wanstreet CE, Calvin J, Overtoom C, Wheaton JE. Bridging the transactional distance gap in online learning environments. Am J Dist Educ 2005;19:105–18.

[40] Christiansen MA. Flipped learning in the broadcast chemistry class [Poster Presentation]; 2016.

[41] Christiansen MA. Flipped learning in the broadcast chemistry class [Oral Presentation]; 2016.

[42] Ferreri SF, O'Connor SK. Redesign of a large lecture course into a small-group learning course. Am J Pharm Educ 2013;77. Article 13.

[43] McLaughlin JE, Roth MT, Glatt DM, Gharkholonarehe N, Davidson CA, Griffin LM, et al. The flipped classroom: a course redesign to foster learning and engagement in a health professions school. Acad Med 2014;89:236–43.

[44] Goldberg H. Considerations for flipping the classroom in medical education. Acad Med 2014;89:696.

[45] Rotellar C, Cain J. Research, perspectives, and recommendations on implementing the flipped classroom. Am J Pharm Educ 2016;80. Article 34.

[46] Jensen JL, Kummer TA, Godoy PD. Improvements from a flipped classroom may simply be the fruits of active learning. Cell Biol Educ 2015;14:1–12.

[47] Rau MA, Kennedy K, Oxtoby L, Bollom M, Moore JW. Unpacking "active learning": a combination of flipped classroom and collaboration support is more effective but collaboration support alone is not. J Chem Educ 2017;94:1406–14.

[48] Shattuck JC. A parallel controlled study of the effectiveness of a partially flipped organic chemistry course on student performance, perceptions, and course completion. J Chem Educ 2016;93:1984–92.

[49] Weaver GC, Sturtevant HG. Design, implementation, and evaluation of a flipped format general chemistry course. J Chem Educ 2015;92:1437–48.

[50] Everly L, Cochran K. Academic outcomes and attitude of pharmacy students regarding flipped classroom teaching in gastroenterology. Am J Pharm Educ 2014;78. Article 111.

[51] Critz CM, Knight D. Using the flipped classroom in graduate nursing education. Nurse Educ 2013;38:210–3.

[52] Findlay-Thompson S, Mombourquette P. Evaluation of a flipped classroom in an undergraduate business course. Bus Educ Accred 2014;6:63–71.

[53] Papadopoulos C, Santiago-Román A, Portela G. Developing and implementing an inverted classroom for engineering statics; 2010. p. TIE1–2.

[54] Prober CG, Heath C. Lecture halls without lecture: a proposal for medical education. N Engl J Med 2012;366:1657–9.

[55] Roach T. Student perceptions toward flipped learning: new methods to increase interaction and active learning in economics. Int Rev Econ 2014;17:74–84.

[56] Smith JD. Student attitudes toward flipping the general chemistry classroom. Chem Educ Res Pract 2013;14:607–14.

[57] Talbert R. How the inverted classroom saves students time. In: Casting out nines; 2011 [Online].

[58] Wong TH, Ip EJ, Lopes I, Rajagopalan V. Pharmacy students' performance and perceptions in a flipped teaching pilot on cardiac arrhythmias. Am J Pharm Educ 2014;78. Article 185.

[59] Robert J, Lewis SE, Oueini R, Mapugay A. Coordinated implementation and evaluation of flipped classes and peer-led team learning in general chemistry. J Chem Educ 2016;93:1993–8.

[60] Utah Education Network Homepage, http://www.uen.org/; 2020. [Accessed July 23, 2020].

[61] Enfield J. Looking at the impact of the flipped classroom model of instruction on undergraduate multimedia students at CSUN. TechTrends 2013;57:14–27. https://doi.org/10.1007/s11528-013-0698-1.

[62] Zydney A, Bennett J, Shahid A, Bauer K. Impact of undergraduate research experience in engineering. J Eng Educ 2002;91:151–7. https://doi.org/10.1002/j.2168-9830.2002.tb00687.x.

[63] Nagda B, Gregerman S, Jonides J, von Hippel W, Lerner J. Undergraduate student-faculty research partnerships affect student retention. Rev Higher Educ 1998;22:55–72.

[64] Christiansen M, Weber J, Sam A, Kingsford T. Positive student responses to embedding a student-chosen research project into a sophomore organic chemistry lab. Chem Educ 2015;20:335–41. https://doi.org/10.1333/s00897152667a.

[65] Christiansen M, Crawford C, Mangum C. Less cookbook and more research! Synthetic efforts toward JBIR-94 and JBIR-125: a student-designed research project in a sophomore organic chemistry lab. Chem Educ 2014;19:28–33. https://doi.org/10.1333/s00897142528.

An accessible method of delivering timely personalized feedback to large student cohorts

9

Cormac Quigley and Etain Kiely

Galway-Mayo Institute of Technology, Galway, Ireland

Introduction

The goal of this project was to deliver meaningful, timely, and instructive feedback to a large group of first-year chemistry students within a resource-limited environment. Here we report on how we used data captured by Moodle and transformed it using MS Office programs to produce feedback that was holistic and timely, offered students insight into their performance and highlighted opportunities for improvement. The output included printed personalized feedback sheets that were delivered to students in a practical/tutorial class setting by a staff member.

This chapter reports how the method of feedback delivery was developed and how it can be adapted to other scenarios. It shows how to create a semiautomated feedback system for students in a large class cohort that is personalized, instructive, and promotes discussion of student performance with academic staff. This system was initially implemented for a cohort of approximately 300 students completing a common first year of a 4-year degree program in Galway-Mayo Institute of Technology in 2016. We report on the creation and implementation of this system and the resultant feedback from students and academic staff on its effectiveness over the past 4 years.

Using a blended learning model, a wide range of information was collected about student participation, grades, interactions, and achievements. The automated feedback system we developed is designed to fit within resource constraints, which means that it can be easily implemented by teaching staff without the need for external help or additional resources. To that end, only ubiquitously available Microsoft office programs such as Excel and Word were used, and data were processed in an intuitive and user-friendly manner.

Feedback and data collection

Feedback is widely recognized as an integral part of a successful learning environment. The importance of feedback in the learning process is hard to overstate. Feedback forms an essential part of the student-lecturer relationship [1], and feedback is, without doubt, necessary for an effective learning environment. This statement comes with a caveat that feedback, if delivered incorrectly, can result in undesirable outcomes, demotivation of students, reduced engagement, and poorer student engagement. Considerable literature exists, which describes the ideal use of feedback, detailing how feedback should be used to set manageable but challenging goals for students to engage with and improve their performance [2]. The use of data as a predictor for directing assistance to specific students in an effort to improve student retention has also been shown to be a worthwhile exercise [3].

In our context, there was a clear need to offer meaningful, personalized feedback to students early in their first year of study. It has also been the experience of the authors that students benefit greatly from early feedback, provided it is accurate, actionable and authentic. In the School of Science and Computing, GMIT, first-year students engage in weekly active learning activities such as participation in lectures, quizzes, practical work, and assignments. Interactions are captured on Moodle (the VLE), which leads to large amounts of data relating to their engagement, time spent, performance, and disengagement being available in digital form. The project sought to turn these data points into meaningful personalized feedback for students.

Although the literature is clear on the value of feedback, providing each student in a large cohort with feedback on all aspects of their performance, particularly in an environment with a high level of continuous assessment, is a demanding task. Indeed, often as cohorts grow in size, individual feedback becomes more fragmented and involves interactions with lecturers, tutors, or teaching assistants who may not be aware of the totality of a student's effort in a given module or subject. Feedback is not always interpreted by the student in the manner intended by the lecturer, rather it is assimilated against a backdrop of their own perception of their progress [4]. On top of this, where students perceive a disconnect between their effort and the success achieved, demotivation is also likely to occur. Such scenarios may be avoided with a teaching environment in which personal interaction with students is an important, integral part of the teaching ethos, but only if the information to provide accurate feedback to the students is also available.

Understanding automated feedback and its limits

Feedback can take many forms; it can be summative or formative, related to a very specific action or process (such as correctly emptying a pipette) or related to performance at a higher level looking at student performance and engagement (such as the number of assignments completed and their average grade).

Feedback on the minutiae of individual actions is not achieved within this feedback system nor is it intended to be. This type of feedback is facilitated during specific assessments and activities, whereas our feedback system is intended to form a holistic view of student performance. This system aims to keep students informed of their progress in coursework at a macroscopic level, looking at how they are progressing, where they have done well, where they could improve, and what their future outcomes could be (with the caveat that past performance is no guarantee of future returns). In this sense, it could be considered that this feedback is a form of learning analytics that are descriptive and diagnostic with a very minor predictive element.

In terms of creating such feedback, often the feedback required on a single piece of student work is simply related to the grade they have achieved. For example, students who achieve different percents will be considered to have passed, passed with some form of commendation, or failed. This varies from assessment to assessment or course to course, but a categorical relationship exists between feedback and achievement. This is quite easily described in a mathematical sense or visually captured using rubrics. Once this is expressed by the lecturer, it can be used to create feedback in an automated fashion. Although this may seem simplistic, and it can be tempting to assume that such feedback would be of limited value to students, it has been our experience that giving the students a global view of their performance and linking their grades to their behavior or level of engagement can be very beneficial to students' understanding of their own progress and trajectory.

More complex feedback that takes into account a greater number of factors can also be expressed mathematically or as a series of logical tests related to numeric values. For example, when considering feedback on combined pieces of work, feedback on an overall performance is contingent on not only the average grade a student achieves but also how they achieve that grade (as well as any critical or must pass components). The feedback for a student achieving 100% in 50% of assignments is different to that of a student achieving 50% in 100% of assignments. This again can be described with only a slight increase in complexity in terms of how feedback is categorized, but this increase in complexity can add considerably to the value of the feedback returned.

Other considerations when designing feedback

Although not unique to this project, it is worth looking at some factors that affect the way in which feedback is designed for this project. The way in which the human mind assimilates information and makes value judgments based on seemingly objective information is subject to a wide array of seemingly irrational behaviors. Indeed, behavioral economics is an entire field of study seeking to explain the irrational and paradoxical choices made by individuals. This is no different when it comes to interpreting feedback [5]. There are many phenomena that are described in behavioral economics, which can be used to optimize the effectiveness of feedback such as loss aversion, sunk cost fallacies, consistency, and (pre)commitment [6]. Although not the focus of the chapter, feedback even when automated, can leverage these. Again, a numerical relationship can be defined to link student performance and the type of feedback that may suitable. A student who is achieving well can be reminded that maintaining their excellent grade can be achieved by continuing their level of work and effort (loss aversion). Students who are struggling because of lack of work can be reminded of their commitment and the value of their work for their intended future career, while improving the chances of improvement by requiring a response indicating their specific intended change in behavior. Other students who are struggling may need to be offered more efficient strategies to tackle coursework such as emphasizing the importance of preparing adequately for practical classes or availing of additional support classes.

Understanding these behavioral biases can motivate learners by using nudge interventions to gently push learners in the right direction. Negative effects may occur if nudges pressurize individuals affecting their intrinsic motivation, so care must be taken to design feedback with an understanding of what behavior the intervention seeks to change. Nudge theory also founded in the behavioral sciences offers a range of intervention types, often classified as passive or active decision-making environments [7]. The first intervention type often referred to as a "true" nudge (e.g., defaults, framing, and peer-group manipulations) targets behavior bias through (small) changes to the decision environment (often subconsciously) without promoting active decision-making. The second intervention type requires specific changes to the decision environment, which includes goal setting, completion tracking, progress bars, and notifications of upcoming deadlines which induce people to utilize these behavioral tools to self-regulate their own behavior [8].

The maths module also adapted a "boost nudge" intervention deliberately aimed at improving learner's decision-making and perception around their maths ability. This intervention focused on teaching students about the limiting behaviors and barriers associated with fixed mindsets. The team ensured feedback promoted growth mindset language, emphasizing "effort" and the importance of making mistakes as part of the learning process.

In any event, when designing a response matrix for different student outcomes, it is worth remaining cognizant of the way in which students may react and respond to the feedback. Although automated feedback alone is not a panacea for improving student learning, it can form a valuable addition to student feedback and certainly can provoke meaningful and effective conversation between staff and students.

Data collection

Collecting student data is considered more fully in Chapter 5. For most lecturers, student data naturally accumulate over the term in the form of grades and attendance. The use of these data to deliver feedback is a normal and expected use of the data, so there are no additional issues regarding data protection legislation when it comes to this feedback system. In any event, lecturers may wish to consult with their local data protection officer and institute policies to make sure that there are no issues. In many cases, the ability to create meaningful feedback can act as a catalyst to improve data collection and storage techniques. It may encourage the storage of data in one single place (such as the VLE) rather than in separate spreadsheets or systems as well as improvements in error correction and timeliness of data collection. Seen in this light, it is a positive development that should be welcomed.

Motivations

The motivation to design this feedback system was the desire to maintain high levels of feedback for students in keeping with the educational ethos of our institute. It was important for us to encourage students and offer an opportunity for them to see how their efforts are contributing to their final marks. It also allowed us an opportunity to flag to students' areas that we would like them to engage with more fully and offer actionable suggestions as to how to improve their overall achievement. In our setting, the first 2 months of the year represent a high risk time for student disengagement, meaning that, for us, suitable feedback in this time frame was likely to be of use in improving student retention.

Comprehensive feedback also provided greater transparency for students who have a very high level of continuous assessment in the first year of their degree. Typically, across all subjects, a student might have three practical classes, three tutorials, and three to six pieces of online work per week, which would each contribute in some way to their final grades. In addition, these grades may be modified by attendance or other factors, which mean that the grading system is nonlinear in nature. Although in practice, this work environment is well signposted and grading schemes are transparent, it is of significant value to student for this to be directly and clearly explained to them at critical points in the academic year. The personalized feedback could also specifically suggest the use of particular services by students, such as the academic writing center or maths learning center, where this was likely to be of benefit.

At the same time, this feedback system had the opportunity to provide lecturers with a more holistic view of students' performance rather than only seeing the portion of their work that directly related to their contact time. This was the case here because our institute ethos is one which cherishes staff-student interaction, and it served as a significant motivation for the creation of the feedback system to facilitate the interaction between staff and student being as valuable to both parties as possible. Lecturers or tutors spend time with the students in all these classes but may never see any other aspect of the student's engagement. As a result, it can be difficult for a lecturer to assess what type of intervention might be suited to a student who is struggling in their class.

It was also seen as a major advantage that this feedback could also be communicated upward to course coordinators, student services, or managers to highlight at-risk students early in their first year of study and facilitate informed discussion with the students. This also lowered the administrative burden on lecturers.

Creating the system

This system is the result of iterative development. It was motivated initially by a desire to make better use of stored data and in essence was an early step toward learning analytics. As it evolved, it also became more complex, improving the amount and type of feedback that could be returned. This changed the motivation from just delivering simple descriptive analytics to one where diagnostic and prescriptive aspects were considered. This also changed the motivation to include a deeper consideration of the way students related to the feedback and the frequency and timing with which it would be delivered. It became clear that it would be advantageous to share the system with colleagues. To achieve this, the system was modified to make it more standardized, and usability was more carefully considered when producing the feedback template files. Importantly, the system, while requiring some know-how to set up initially, requires no specialist software or training and can be used by anyone reasonably adept at using MS office software applications.

Moving from data to personalized feedback

In this case study, and the examples shown, data were obtained from Moodle 3.6 and earlier Moodle versions. The process is, however, easily adapted to any data source and is entirely VLE agnostic. The process of converting the data into useful feedback has three stages. These steps are achieved using accessible tools such as MS Excel *(steps 1 and 2)* and MS *Word (step 3)*.

1. Extracting the data from the source and cleaning the data.
2. Designing feedback and applying the automated feedback algorithm.
3. Generating the feedback sheets.

Step one: Target data are extracted and curated

Typically for a feedback sheet, gradebooks and attendance records are downloaded from Moodle. It is possible to include any data source such as separately recorded results. In this case, Moodle stores these as separate tables that can be downloaded as spreadsheet files (.xlsx or .csv etc.). These worksheets are combined into one single worksheet containing all the source data. In this project, the Moodle gradebook is usually set up to have grades processed to the greatest extent possible so that less manipulation of data is required. In all cases, when combining worksheets, it is essential that the rows in both spreadsheets are in identical order. (This may seem an obvious point, but it is worth emphasizing.) Rather than manually checking each row, an automated check can be run to identify any rows for which rows do not match. Using an IF statement such as $=if(Column1 = Column2, 1, 0) + if($(for as many columns are needed to ensure matches are unique) followed by a count in the title row of $=Countif(Column, = 0)$ will count how many are not matched.

Once the data have been combined in a single source, and depending on the source, it will be required to be curated. There may be extraneous information, partial or inaccurate scores, or grades that must be combined in a particular manner. Before moving onto step 2, your spread sheet should contain all the numerical data you wish to report on. This might include nongrade items such as marks lost to absence, hypothetical grades without penalties applied, distance to significant grade levels, or any other relevant data.

For the chemistry and maths modules, which this case study is based on, the grading system contains a number of nonlinear features. Grade limiters are used (for failure to complete prelaboratory and postlaboratory work), attendance penalties are enforced, and students are given a worst grade discard. In early iterations, each of these was completed in Excel; however, Moodle gradebook (using the advanced formula feature) now allows these to be calculated within the gradebook, which greatly simplifies the processing of the worksheet.

Step two: Design the feedback matrix and transform data

In short, once the data reflect the assessment strategy with correct values/grades for the required items, a set of responses for each item must be generated. Once the feedback has been written, it is placed in the corresponding column by using a set of nested IF statements, before exporting the feedback in step 3. Creation of this feedback requires careful attention and is the most time-consuming step in the process; once complete, however, it can be reused in future iterations of a course or adapted to other situations.

In practice, this is achieved by the creation of a feedback matrix, which is stored as a separate worksheet within the Excel spreadsheet. From here, it can easily be referenced by formulae in the separate worksheet containing the student data but modified without fear of introducing errors into the student data. It can also be easily copied into another spreadsheet for future use. An example of a feedback matrix is shown in Fig. 9.1 where the grades and the corresponding feedback can be seen. This example is from the original implementation of the feedback sheets. The creation of a matrix like this allows the feedback designer to quickly manipulate each of the different feedback components and the corresponding criteria. In the original form, entire "IF" statements were written with all feedback inside; however, this made later editing *very* difficult. Each line contains the feedback merited by the corresponding grade category indicated on the far side. The numbers correspond to the minimum numeric grade required to receive that feedback. There are also columns for specific exceptions allowed, which can be indicated by strings of characters.

It is the intent that the chosen feedback phrases will form part of a letter, which in all cases should read in a natural and free-flowing manner. It is, therefore, important that the different feedback sentences are carefully structured to ensure this flow. Each sentence should have the same grammatical construction. Aside from this, all the usual considerations should be given to how the feedback will be received by students, and a balance must be struck between admonishment and encouragement, but this is a matter of lecturers' prerogative.

The example shown in Fig. 9.1 is an early matrix used in this project with a limited set of five responses for each factor, that is, accumulated weekly quiz performance, attendance, laboratory grades, assessment grades, and stand-alone quiz performance.

This can be expanded in a number of different ways. By creating a hypothetical grade that is proportionately increased as if the student had 100% attendance,

Comments for:	Perfect score	Positive	Medium	Poor	Zero	Missing	Perfect score (score >= this)	Positive (score > below but less than perfect)	Medium (score > below but less than positive)	Poor (score <medium but > 0)	Zero (score <0, non attendance comment)	Excused/resit required/other
Lab grade	You lab work is going very well, keep up the good work.	Your lab work is going well, keep it up. Don't forget to read your lab book before labs.	You are passing your labs but your performance could be improved. Remember to read your lab book before you come to labs.	It looks like you are struggling with the labs. If you feel you need extra help come talk to me.	You haven't gotten any grades yet, please talk to me immediately.			9	8	6		0ex
Lab attendance	You have attended every lab so far, well done.	Keep up the good attendance at labs.	Your lab attendance needs to improve. Remember there is an 80% attendance minimum to pass the labs. You must pass the chemistry labs to progress to second year.	You have missed a large number of labs. If I have not already spoken to you about this come see me.	You haven't been recorded as attending a single lab class this term, please talk to me immediately.		100	80	60			0ex
Test grade	Well done on a perfect score!	Well done on the great score on the chemistry test. Remember to keep on top of your chemistry work as the year goes on.	Well done on passing the chemistry test. You should aim to improve your marks for the next one though, keep working at it.	You failed the chemistry test. The standard for these tests is the standard required to pass the year so now is the time to turn this around. You will need to put in much more work.		0You missed the chemistry test.	100	70	40		0	absent
Quiz grade	Well done on a perfect score!	Well done on the great score on the chemistry quiz. Calculations are important for success in the lab.	Your quiz score could be improved, you should revisit the practice quizzes to improve your skills. Improving your skills will also improve your lab performance.	You need to improve your quiz score, ask if you need extra help. You will need to put more time into practicing your calculations. They are essential for being able to pass the lab.	You did not do the chemistry quiz, you have forfeit these marks. Make sure you can do all the questions in the practise quizzes. The calculation skills are essential in the lab.	You did not do the chemistry quiz, you have forfeit these marks. Make sure you can do all the questions in the practise quizzes. The calculation skills are essential in the lab.	15	12	6		0-	
Final comment	Keep up the good work!	You are passing comfortably but with a few changes, your grades could improve significantly.	If you are struggling talk to your lecturers in the lab, it is a lot easier to change things now then 4 weeks before the exams!	If you are struggling talk to your lecturers in the lab, it is a lot easier to change things now then 4 weeks before the exams!			70	55	40		0	

FIG. 9.1

A feedback matrix. A feedback matrix showing a limited set of feedback criteria and the corresponding feedback, which would be returned to the student.

feedback can demonstrate marks lost and, as importantly, improvements in future grades, which can be gained by behavioral changes. This is included only if greater than a certain threshold because it is not relevant to all students, so there may be blank cells within the feedback matrix or there may be several variations of a single feedback item. Similarly, with marks lost to failure to complete prelaboratory quizzes, a hypothetical grade can be offered only if it is relevant. Finally, the letter sign off can be selected as a response to any combination of factors and creates further personalization of the feedback.

When designing the feedback grid, it may seem that only five or six different lines of feedback per item is too few; however, the variety offered by even five feedback lines per item when combined over several items means that the total number of possible variants of the overall feedback quickly becomes very large. For a simple matrix containing five options on five items, 3125 outputs are possible.

A short note on the "IF" function

The use of the IF function within Excel is crucial to the operation of this method of feedback delivery. It should be said that this is neither the most computationally efficient nor elegant method of completing this; it is, however, relatively simple, ubiquitously available, easily edited and easily shared. The IF function has the following syntax =If(logical test, true, false). It performs a logical test and then returns the first value after the comma if it is true; otherwise, it returns the second. It can be nested by placing a further IF statement as the value returned for either the true or false value. The nested IF statement is processed in a sequential manner with the outermost test being conducted first.

By way of example, examine the following statement:

$=$ if(grade > 40,"passed",if(grade > 50,"merit","fail")) where grade is the student score being evaluated.

Here, first we test whether the grade is greater than 40; if this is true, then "passed" is returned; if not, it proceeds to check whether the grade is greater than 50. This, of course, will not happen because if it is greater than 50, it will also have been greater than 40, so the first statement will return "passed" rather than running this test. It can be worth drawing out a tree to describe the sequencing of the logic if you are attempting to produce a particularly complicated example. In addition, for a value of 40, the function will return fail because the grade is not greater than 40. Usually for continuous variables, a greater or less than test is used; otherwise for categorical variables = is used. Excel allows for up to 64 levels of nesting, which means that it will be possible to create even the most complex of sorting algorithms.

A typical "IF" statement is shown in Fig. 9.2. The cell references in red refer to the cell containing the student data, whereas the highlighted cell references correspond to the feedback matrix. The feedback can be taken from the relevant cell, and the IF statement can be changed using the find and replace facility rather than by producing a new statement each time. Each of these requires the feedback matrix to follow an identical format on each row but is advantageous because each

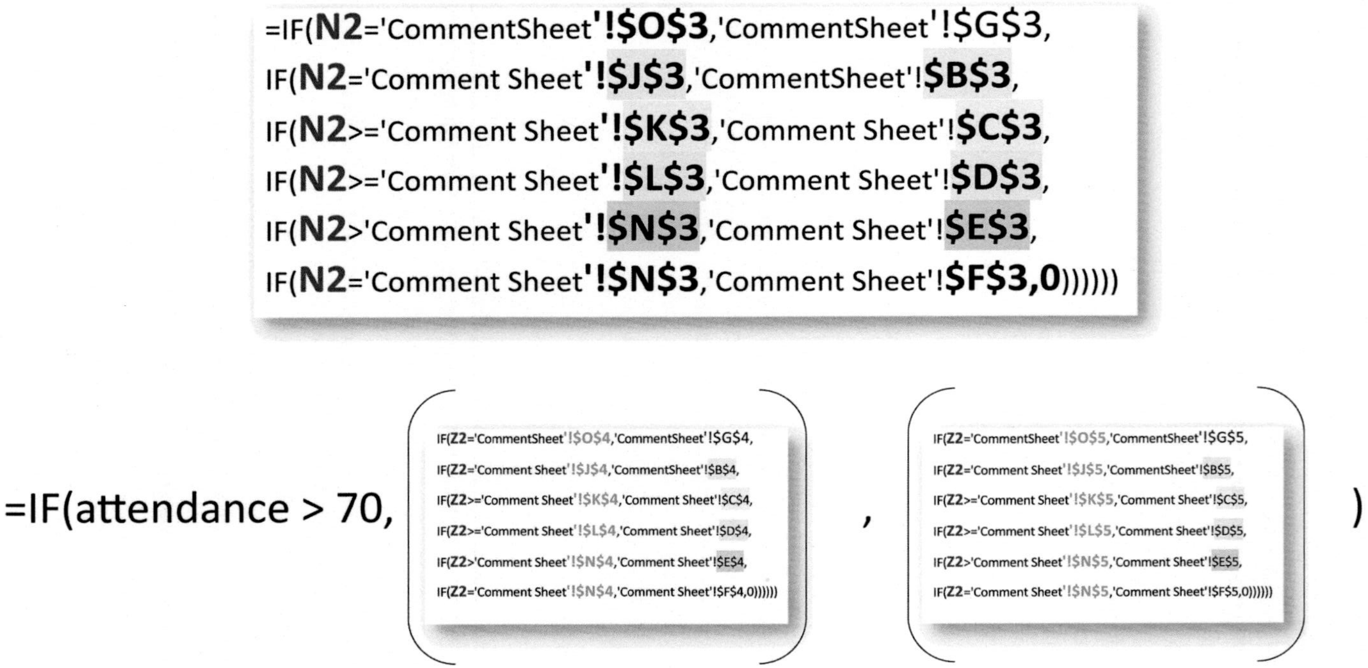

=IF(**N2**='CommentSheet'!$O$3,'CommentSheet'!$G$3,

IF(**N2**='Comment Sheet'!$J$3,'CommentSheet'!$B$3,

IF(**N2**>='Comment Sheet'!$K$3,'Comment Sheet'!$C$3,

IF(**N2**>='Comment Sheet'!$L$3,'Comment Sheet'!$D$3,

IF(**N2**>'Comment Sheet'!$N$3,'Comment Sheet'!$E$3,

IF(**N2**='Comment Sheet'!$N$3,'Comment Sheet'!$F$3,0)))))

=IF(attendance > 70,

IF(Z2='CommentSheet'!$O$4,'CommentSheet'!$G$4,
IF(Z2='Comment Sheet'!$J$4,'CommentSheet'!$B$4,
IF(Z2>='Comment Sheet'!$K$4,'Comment Sheet'!$C$4,
IF(Z2>='Comment Sheet'!$L$4,'Comment Sheet'!$D$4,
IF(Z2>'Comment Sheet'!$N$4,'Comment Sheet'!$E$4,
IF(Z2='Comment Sheet'!$N$4,'Comment Sheet'!$F$4,0)))))

,

IF(Z2='CommentSheet'!$O$5,'CommentSheet'!$G$5,
IF(Z2='Comment Sheet'!$J$5,'CommentSheet'!$B$5,
IF(Z2>='Comment Sheet'!$K$5,'Comment Sheet'!$C$5,
IF(Z2>='Comment Sheet'!$L$5,'Comment Sheet'!$D$5,
IF(Z2>'Comment Sheet'!$N$5,'Comment Sheet'!$E$5,
IF(Z2='Comment Sheet'!$N$5,'Comment Sheet'!$F$5,0)))))

)

FIG. 9.2

Nested IF statements. A typical nested IF statement used to correlate the feedback matrix with the student results. The lower statement shows how these can be combined to increase complexity without increasing the complexity of the feedback matrix.

statement will change only in the referenced cells. In addition, editing feedback when it is stored in a standardized format is considerably easier because edits to each cell will be reflected in the feedback assigned and will update itself any time a change is made.

It is also possible to examine multiple factors to provide feedback to distinguish between a student with high attendance and low marks compared with a student with high marks but infrequent attendance. Here, the outermost IF function can be used to direct the feedback to the appropriate line on the feedback matrix. Again, by using the standard matrix format, this statement can be edited easily, and the complexity of the feedback matrix is not increased. Fig. 9.2 shows a theoretical representation of this. In addition, note that it may not be necessary to fill in all squares in the grid, for example, if the cut off for a certain feedback category is 70% attendance, then there is no need to fill feedback for grades higher than 70% in that line of the feedback matrix.

Step three: Generating and distributing personalized feedback forms

Once the spreadsheet is populated with the grades and responses, all that remains is to use the mail merge function in Microsoft Word to produce the final document for sharing. This is by far the simplest part of the process. Examples of these are shown in Fig. 9.3. With a small amount of trial and error, the sheets can be reliably generated very quickly, particularly if care is paid when creating the feedback to ensure consistency of syntax, use of punctuation, etc. The sheets can be distributed either (as in the case of this project) by hard copy or alternatively by email or sent in chat during a 1-1 MS Teams video call. Importantly, by tailoring the sentences used as feedback, it is possible to create a feedback letter that appears completely personalized to the student who receives it. In the first iteration of the feedback sheet, it reviewed five aspects with five possible responses using a simple matching function. This having 3125 possible outputs, meant that, even at this relatively low level of complexity, most students were unlikely to receive identical feedback to their peers.

In subsequent versions, additional aspects of student performance were reviewed, and two extra features were added. With the addition of these feedback features, there are over 1 million permutations possible. The sheets provided actionable suggestions for students to change how they engage with the module, with the aim of producing long-term positive changes in student behavior. Feedback is distributed to students by lecturers or tutors and sent to course coordinators for reference as needed.

A key part of the effectiveness of the feedback is how it is delivered to the student. In this case, students reviewed this personalized feedback sheet with their lecturer or teaching assistant during tutorial or practical class time. This provides an opportunity for students to clarify any uncertainty around the content of the feedback sheet, ensuring that the feedback is meaningful. It also creates an opportunity to ask questions and discuss learning strategies or interventions that were suggested in the feedback.

Maths 1.1 Personalised Feedback Form

Group

Dear student

We hope you are enjoying first year maths. Here is a summary of your progress in maths so far.

Journal Work (20%) (out of 5%)	Quizzes (15%) (out of 4.5%)	Exams (65%) Week 7 (out of 10%)	Marks so far (out of 19.5%)	Average Mark so far %
3.3%	1.5%	7.0%	11.8%	60.7%

Remember there is no such thing as a maths gene. All our brains have a remarkable capacity to grow and change and with **frequent practice and effort.**

Lecture Class Participation
Your participation at lectures: **100.0%**
You are attending most of your lectures, good effort.

Journal Work (Mandatory Attendance)
Your participation in journals is **100.0%**
Great effort on attending all your journal classes.

Your mark for Journal 1 and 2 is averaging at 66.7%. Because your journal mark is weighted by attendance your mark becomes **66.7%.** This effort contributes towards **3.3%** out of 5% available so far for journal work.

Moodle Quiz Effort

Of the 3 semester 1 quizzes so far, you have achieved 100%, mastery in 1.0. This equates to **1.5%** of the 4.5% available for quizzes. *You have achieved a 10/10 well done on trying. Keep going until you achieve 10 on all quizzes, they are a great way of learning through retrieval and practice and build strong neural pathways.*

Only the quizzes that you get 100% in will **contribute towards your final mark.** When we make a mistake, synapses fire in your brain which means learning occurs. Quizzes are a great opportunity to practice and learn from your mistakes and are used again in exam questions.

Week 7 Exam

You achieved **70.0%** the week 7 exam. This contributes to **7.0%** of overall marks in maths.

Please feel free to ask questions in lectures or in journal classes.

The **Maths Learning Centre** offers **free maths** tuition to GMIT students located in the library training room 973. It is a drop-in service, and **no booking is required**. Find the times at

 https://library.gmit.ie/support/math-learning-centre/

Please chat with the maths team if there are any inaccuracies in what is presented here, and we can review it for you before you sign it.

Student Signature (*Please sign this when you agree with data presented here*)

Total Assessments =	10 quizzes (15%)	8 Journals (20%)	5 Exams (65%)

FIG. 9.3

Examples of feedback sheets showing different possible styles of feedback to hypothetical students. The example on the left shows a typical Maths feedback sheet, and the right shows a typical Chemistry feedback sheet.

School of Science and Computing

Amedeo Avogadro   Group A   G00602214@gmit.ie

SFSCG_H08_Y1

Hi Amedeo,

I hope term one is going well for you. As we approach the end of this term it is time for a quick recap on how you are getting on with Chemistry.

- **Starting with attendance**, of the first 9 labs, you attended 8. This gives you an attendance of 89%. Keep up the good attendance at labs.

- Your **grade for the labs** over these nine labs was 58.6%. You are passing your labs but your performance could be improved, lets aim to do this in term 2. Don't forget the prelab quiz and to look over your lab book before lab each week.

- **You completed 11 quizzes** of a possible 16 so far this term. You missed a few quizzes, you should try and keep on top of them. The quizzes you missed reduced your overall lab mark by 1.1%. If you complete all of the quizzes next term this number will be reduced to zero. Your average quiz score was 82.1% If you had completed all the quizzes you would have improved by 6.8 and your lab mark would have been 65.4%.

- Looking specifically at the **Moodle Midterm Quiz** lab, you got 0 out of 10. You did not do week 8 Moodle lab, you have forfeit these marks. Make sure you can do all the questions in the practise quizzes. The calculation skills are essential in the lab.

- Your **actual contribution towards your end of year grade** from the first nine practical classes and the midterm quiz is 6.6% out of a possible 12.5%. You are passing the practical component of this module so far. You should improve your grades next term. The practical skills you gain in the coming term will be important for the next three years (and beyond) no matter your subject choice.

- In your **theory assessment** last term, you scored 86%. Well done on the great score on the chemistry test. Remember to keep on top of your chemistry work as the year goes on - the next exam will be more of a challenge!

- The **Practical exam** is next week and, as you will have seen from the marking scheme, it will test your accuracy, precision and calculations. It is worth 10% of your total mark for chemistry this year so it is worth preparing for.

As always, if you have any questions feel free to ask. Best of luck with the exam next week.

Findings and conclusion

This system has been in use since 2016 and has been well received by both students and staff. Since then, it has been adapted and adopted by lecturers in other departments within the institute and shared through the Irish National Forum for the Enhancement of Teaching and Learning in Higher Education ORLA project [9]. Over this time, student surveys have been used to collect feedback from students midway through and at the end of the academic years from 2016 to 2020 to assess the impact of the system in Maths and Chemistry for first year undergraduates.

Feedback from students

When looking for feedback from students, we were interested in whether the feedback sheet had any positive effects on the students. By its nature, it would be almost impossible to observe a quantitative change in student behavior because of the feedback sheets; however, the use of surveys to qualitatively investigate student responses to the feedback was fruitful.

The first questions about the feedback sheet that we were interested in answering was whether or not students found the sheet useful and if so why. Beyond this, what was of interest was whether or not it had any effect on the students' future plans, and if on reflection at the end of the year, they could relate any changes they made in their behavior to the feedback they received. Students were asked in the survey to identify any changes they intended to make at the middle of the year and any changes they did make at the end of the year. This was captured using short surveys containing Likert objects and free response questions. The students have been consistently surveyed over 3 academic years in both their Maths and Chemistry modules.

Each year, an overwhelming number respond positively, either agreeing or strongly agreeing with the statement that they find the personalized feedback sheet to be useful. In 2019, 94% ($n = 229$) of Maths students and 95% ($n = 162$) of Chemistry students returning the survey agreed with this. Similar findings were reported in 2020 where 94% ($n = 303$) of students strongly agreed or agreed that the personalized feedback form was useful.

Impact on students

As a single statement, finding the form useful is limited in what it can tell us. It was initially surprising to the authors that the feedback was equally useful to high-performing students. There was no statistically significant difference in grade when grouped by how useful students reported finding the feedback. Thematic analysis of the free responses allowed us to better understand how the form was of use to different students with different situations. Two main themes emerged, one from students who were doing well but uncertain and a second from students who were not engaging with the course at the level that they might aspire to.

Within the first theme, what came through from students was that feedback provided reassurance that their learning was going well. Many students reported less anxiety or increased confidence after receiving the feedback, as the two examples from surveys below demonstrate:

I received a lot of feedback regarding my performance and grades which was helpful in knowing how well I was doing in the module. It made me less anxious to know I was doing well.

Student 1

I felt it was good to see my progress early on in the year, it gave me more confidence that I was on the right track.

Student 2

At the same time, the feedback, because it is personalized and may appear completely different for students who are struggling in different areas, was met with different responses for these students. In this case, the student's responses focused on the benefit of seeing opportunities for improvement. Again, the feedback below was typical of the responses from students whose responses aligned with this theme.

The personal feedback forms were very useful as they keep you up to date with your progress and it also helps you to see what you can do to improve your grade.

Student 3

Impact on learning

Most learners responded that the personalized feedback would change their approach to studying that subject. This was examined in Maths and Chemistry and consistently around 70% of students report that they would make changes. Learners indicated that the personalized feedback would change their approach to learning Maths and Chemistry in 2019. This was 72 % ($n=159$) and 68% ($n=229$) for Chemistry and Maths, respectively. In 2020 again 74% ($n=303$) of Maths students responded that they would change their approach.

Students were asked to give an example of a change/practice they were likely to make as a result of receiving the feedback. Fig. 9.4 shows how students responded when offered different suggestions. This is also reflected in their free response answers; for some students, it was a motivating factor to continue working hard. For others, they could identify specific changes to their behavior, which would improve their learning. In both cases, the feedback was observed to positively affect learners.

Well I will personally keep working hard to try keep getting really good marks, I was never good at maths but hard work and persistence will pay off in the end.

Student 4

I am going to focus more on the quizzes and aim to get 100% weekly and also I am more encouraged to challenge myself in the quizzes.

Student 5

What choices/changes would you "make" as a result of your personalized feedback summary?

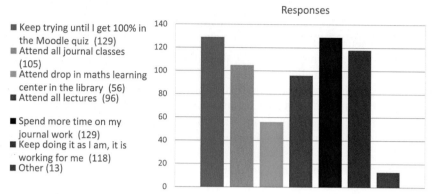

- Keep trying until I get 100% in the Moodle quiz (129)
- Attend all journal classes (105)
- Attend drop in maths learning center in the library (56)
- Attend all lectures (96)

- Spend more time on my journal work (129)
- Keep doing it as I am, it is working for me (118)
- Other (13)

FIG. 9.4

Student response to suggested changes in behavior prompted by automated feedback. Students responding to the survey indicated a variety of proposed changes to their study practices.

> *I will keep going over semester 1 quizzes as this will keep the material in my head also, I will keep attending the lectures and tutorials. I will keep my journal work up to a high standard.*
>
> **Student 6**

Feedback from lecturing staff and management

At the same time, because the feedback is distributed on a one-to-one basis to students by a relatively large number of staff members (>15), it has been possible to get feedback from lecturers. This has also been positive with lecturers giving feedback such as:

> *Student performance is ready at a glance and facilitates an honest one on one discussion.*

> *It gives an overview where the lecturer normally only sees one aspect of student performance.*

These both reflect the fact that the holistic nature of the feedback allows both staff and students to see how students are progressing and then make choices on an informed basis. This has meant that the feedback is now targeted to provide students with feedback and encouragement at times of the year where dropout rates are known to be at their highest level. This aims to encourage students to re-engage with their program, improving student outcomes by empowering students to make active, informed decisions about their engagement with their course and keeping as many students as possible on track to achieving their academic potential.

Lessons learned

Because this system has been in use and grown over the past 5 years, there have been a number of lessons learned. During that time, learning analytics has gone from a niche interest to something that is a topic of conversation in nearly all higher education institutions. One of the most important realizations was that what we are undertaking with this feedback system is learning analytics and that descriptive learning analytics can be a powerful way to interact with students and improve their educational experience. Moreover, this can be done with what is, by the standards of a commercial analytics project, a tiny amount of data and computational resources. More important than a large quantity of data was high quality of data, a commitment to keeping timely and accurate information enabled the project.

For anyone considering implementing a similar feedback system, I would offer a few thoughts. Data analytics is no substitute for high-quality teaching. How the results of the data analysis are shared is as important as the analysis itself. Consider how students will be given any feedback—will they have an opportunity to query any results? Will the meaning of each output be clear or require clarification? Might the outputs be misleading or demotivating?

The transparency of the system has been well received by students because it also offers an error correction mechanism for any perceived inaccuracy in stored grades, empowering students by allowing opportunity for discussion. Early timing of the feedback offers students an opportunity to make changes within the first semester, and there is a balance to be found between early feedback and comprehensive feedback. Typically returning feedback as soon as there is sufficient data has been beneficial. In our case, we try to ensure that feedback forms are produced within the first 8 weeks of term to both motivate students and offer them an opportunity to change behavior with enough time to see the effect of their changes. We have found that the feedback elicits a much stronger and more well-informed response from the student regarding assessments compared to marking feedback on individual assessments. The modules in which this has been used have changed in tandem with the implementation of this feedback system rather than changing because of it, but it has been an avenue for evaluating student achievement.

Finally, in creating this system, it has become apparent that a large amount of data about each student's performance is available to each lecturer. By taking stock of the data that are available and improving the quality of what is there rather than looking for new resources, it is possible to provide a useful service to the students. Implementing this system can be done with no additional digital resources; while there was a big initial investment of time in setting up the system, it has now become a time saver and offers a readily accessible way to offer improved feedback to large cohorts of students. Hopefully, this chapter will shorten the path for others.

References

[1] Hattie J, Timperley H. The power of feedback. Rev Educ Res 2007;77:81–112. https://doi.org/10.3102/003465430298487.

[2] Marshall S, Fry H, Ketteridge S. A handbook for teaching and learning in higher education: enhancing academic practice. Taylor & Francis; 2014.

[3] Pardo A, Jovanovic J, Dawson S, Gašević D, Mirriahi N. Using learning analytics to scale the provision of personalised feedback. Br J Educ Technol 2019;50:128–38. https://doi.org/10.1111/bjet.12592.

[4] O'Donovan BM, den Outer B, Price M, Lloyd A. What makes good feedback good?. Stud High Educ 2019;1–12. https://doi.org/10.1080/03075079.2019.1630812.

[5] Brown GTL, Peterson ER, Yao ES. Student conceptions of feedback: impact on self-regulation, self-efficacy, and academic achievement. Br J Educ Psychol 2016;86:606–29. https://doi.org/10.1111/bjep.12126.

[6] Cialdini RB. Influence: science and practice. Pearson Education; 2009.

[7] Damgaard MT, Nielsen HS, et al. Econ Educ Rev 2018;64:313–42. https://doi.org/10.1016/j.econedurev.2018.03.008.

[8] Grüne-Yanoff T, Hertwig R. Nudge versus boost: how coherent are policy and theory?. Mind Mach 2016;26:149–83. https://doi.org/10.1007/s11023-015-9367-9.

[9] Quigley C, Kiely E. Harnessing student engagement data for personalised feedback, https://www.teachingandlearning.ie/resource/harnessing-student-engagement-data-for-personalised-feedback/; 2020. [Accessed July 30, 2020].

Applying NuPOV to support students' three-dimensional visualization skills

10

Jia Yi Han[a], John Yap[b], Teck Kiang Tan[c], Yulin Lam[a], and Fun Man Fung[a,c]

[a]*Department of Chemistry, National University of Singapore, Singapore, Singapore*
[b]*Application Architecture and Technology, NUS Information Technology, Singapore, Singapore*
[c]*Institute for Applied Learning Sciences and Educational Technology (ALSET), NUS, Singapore, Singapore*

Introduction

In recent years, augmented reality (AR) and virtual reality (VR) have been widely discussed in technology. As technology advances rapidly over the past decades, it is not surprising to see that VR and AR technologies have embedded deeply into major industries globally [1–3]. Education is one of the sectors in which many countries have sought to develop continuously [4, 5]. As demand for employees possessing higher education increased globally [6], innovative educators and novel education techniques have also been increasingly sought after [7–9]. Scientific subjects such as physics and chemistry are known to involve not only sophisticated mathematical calculations but also consist of concepts that require visualization of spatial elements [10–12]. As these concepts are usually conveyed using two-dimensional media (i.e., pen and paper and computer screens) by many educational institutions [13–16], it is understandable that educators may find spatial visualization concepts challenging to teach to students and assessing the students' understanding of them. Considering these difficulties faced by students and educators, three-dimensional media could be used in higher chemistry education, and VR and AR might be a solution to this problem [17].

There have been various attempts at incorporating VR and AR technologies into education [18–21]. Studies have shown how VR and AR technologies have seen usage in many higher educational institutes globally, being made accessible and affordable to the layperson because of the smartphone's ubiquity in the modern era [22]. Furthermore, specifically in the field of chemistry education, there are attempts by educators to integrate AR in chemistry educations such as the development of mobile applications such as "Elements 4D" and "Chirality-2" [23, 24].

Technology-Enabled Blended Learning Experiences for Chemistry Education and Outreach
https://doi.org/10.1016/B978-0-12-822879-1.00002-0

The NUS-developed "3D Sym Op." [25], while not AR/VR, was a mobile application that aimed to help users visualize to identify a molecule's symmetry operators and point groups by helping users to visualize the molecules. By allowing users to view a molecular model on their mobile device, it enables them to figure out for themselves why certain molecules are of the point groups they are. The vision of our project is quite similarly inspired in that, through AR technology, we aim to enable students to gauge their understanding on reaction mechanisms.

The current tools available for visualizing chemical reactions are largely limited to static models and preprogrammed animations [26]. Although those tools could show users the chemical reactions process, the lack of user interactivity omits a crucial aspect of teaching; students would find it challenging to visualize changes to these models arising from reaction mechanisms. Developing such cognitive skill—the spatial visualization skill—is essential in understanding chemistry.

Nucleophile's Point of View (NuPOV) uses AR technology to assist in teaching chemistry concepts to freshmen undergraduates [27]. It does not claim to be able to replace conventional teaching methods with our project. Conventional lectures and written tests are still needed for students to assimilate the concepts [28]. However, it was hoped for NuPOV will act as a more advantageous supplementary tool to be used with conventional teaching methods compared with other supplementary tools available in the market. Our application enables students to improve their spatial understanding on how molecules undergo changes to their structure as the nucleophilic addition reaction progresses and the mechanism behind the said reaction.

The mobile application

Spatial visualization, one of the many cognitive skills required in learning chemistry concepts, plays a pivotal role in establishing good fundamentals in understanding chemical reactions. NuPOV aims to help chemistry students improve their spatial visualization ability (SVA) for chemical reactions by showing them how reactions occur in a three-dimensional space. In our prototype, nucleophilic addition was chosen as the topic for visualization because it is one of the organic chemistry's most fundamental concepts [29] (Fig. 10.1).

A partnership was formed between the project team from the Department of Chemistry and education technologists and programmers from NUS IT to assist with the development and coding of the mobile application. Under the guidance and vision of members from the project team, the development of "NuPOV" mobile application's prototype was completed.

In NuPOV, by "firing" a nucleophile at a molecule, users can simulate the nucleophilic addition reaction. In the process, they need to fulfil the specific conditions as stipulated by the collision theory for a successful reaction to occur [30]. This mechanistic approach not only enables users to observe how the molecules undergo changes during the reaction but also to learn about the different physical and spatial factors at play, which could affect reaction through interacting freely with the molecule. By allowing students to clarify their own doubts on the concepts of nucleophilic addition through individualized AR interactive experiences, self-directed learning is achieved [31, 32].

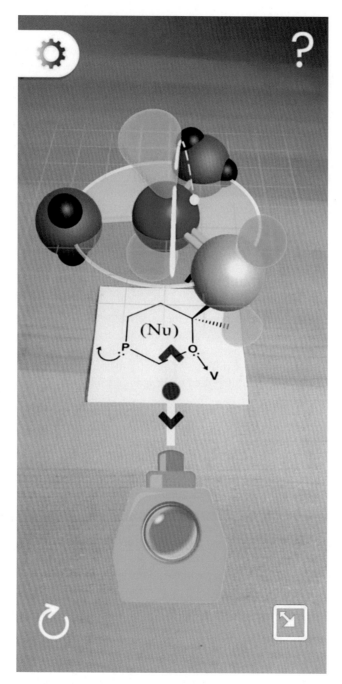

FIG. 10.1

Screenshot of the "NuPOV" mobile application.

To our knowledge, NuPOV is the first mobile application to allow users to spatially interact with molecules in a reaction. Similar to other visualization aids, one can learn what happens to the molecules during the reaction. However, with the added depth perception of the AR element, as well as the interactivity of the process, students are also able to spatially visualize when and where each steps of the reaction mechanism occur between the reactants.

Despite its minimal appearance, the prototype version of NuPOV already serves most of the functions the planned full version is to encompass. The only limitation that it has is in with the limited number of molecules that can be interacted. The prototype teaches users the basic visualization skills required to understand the nucleophilic addition reaction mechanism for simple molecules. Once more molecules are available in NuPOV, users will not only be able to grasp the reaction concept for more complex molecules, but they will also have better understanding of the spatial aspects of other reaction mechanisms such as nucleophilic substitution, cycloaddition, or even catalytic reactions.

The experimental framework

We aim to explore if a higher SVA is associated with learners demonstrating greater chemical competency. Thus to this end, the experiment was planned to adopt a pretest/posttest design that captured changes in the students' SVA [33, 34].

Given that SVA is quite a large concept that encompasses a wide variety of factors and skills, it would prove difficult to directly measure it. However, a study by Tasker and Dalton [26] had identified three main attributes that determined SVA in students:

- "Disembedding ability"—the ability to identify crucial details in visual displays.
- Visuospatial memory capacity.
- The ability to relate old information with new information.

These main attributes were used to measure the students' SVA both before and after interacting with the mobile application to measure a change and, hence, its effectiveness (Fig. 10.2).

We had also accounted for several other factors that may indirectly influence SVA and chemical competency apart from the effect of our mobile application [35]. Attributes influencing SVA included demographic information gender and age [36], whereas the attributes influencing chemical competency included self-efficacy [37], aptitude for learning [38, 39], and interest in chemistry [40–42]. These factors also had to be measured as control variables in our experimental design to ascertain whether the students' change in SVA throughout was directly caused by our mobile application. It was hypothesized that most of the control variables would remain constant throughout the course of the data collection.

However, given that the mobile application was still in quite a preliminary stage and only dealt specifically with the chemistry concept of nucleophilic addition, we anticipated that the application would not be able to help students improve their overall SVA as it was. Instead, a trial of a smaller scale had to be executed to determine

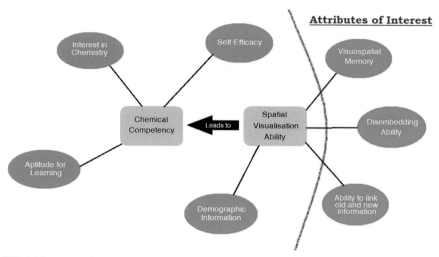

FIG. 10.2

List of miscellaneous factors that might influence spatial visualization ability (SVA) and chemical competency. Our attributes of interest listed on the right.

whether the mobile application helped students understand the spatial implications of nucleophilic addition. That data would then be extrapolated to infer whether they would help students understand the spatial elements of other chemical reactions if they were to be implemented into the mobile application in future. Such a trial was integrated into the curriculum of a first-year organic chemistry undergraduate module because it was a module that included the concept of nucleophilic addition.

Results

Measurement of control variables

The pretrial survey aimed to obtain data on the control variables through inferring from the main data. Thus the results from this survey are only significant with the data collected in the posttrial survey. Similar questions were used in both surveys, and each question tackled one of the three more malleable control variables (self-efficacy, learning aptitude, and interest in chemistry). Results from the two surveys were compared against each other to check whether there had been any significant change in the control variables (Table 10.1).

The comparison took responses from a sample of 87 students who sat for both the pretrial and posttrial surveys and excluded an attrition of 45 students who attended the pretrial but not the posttrial. The comparison has shown that p-value <1 in the control factors affecting SVA and competency in chemistry among the students who attended the trial. This meant that any measured changes in SVA (regarding nucleophilic addition in particular) could indeed mostly be attributed to their interaction with the mobile application rather than the other miscellaneous factors.

Table 10.1 Comparing control variable results measured from pretrial and posttrial tests, together with question topics corresponding to each variable.

Measured Variable	Question Topic	Change in Score									Average Change (+/-)
		-4	-3	-2	-1	0	1	2	3	4	
Self-Efficacy	Confidence in their own understanding	0	0	0	10	38	28	7	4	0	0.50575
	Confidence in solving difficult questions	0	0	0	5	24	37	17	3	1	0.90805
	Confidence in teaching peers	0	0	1	8	38	30	7	3	0	0.49425
Learning Aptitude	Not requiring help in topic	0	0	1	4	20	37	15	10	0	1.04598
	Willing to take up difficult mods	0	0	8	27	24	21	6	1	0	0.08046
Interest	Enjoy CM1121	0	0	7	20	35	20	4	1	0	0.03448

Reproduced with permission from Fung et al., Higher Education Campus Conference (e-HECC), 2020.

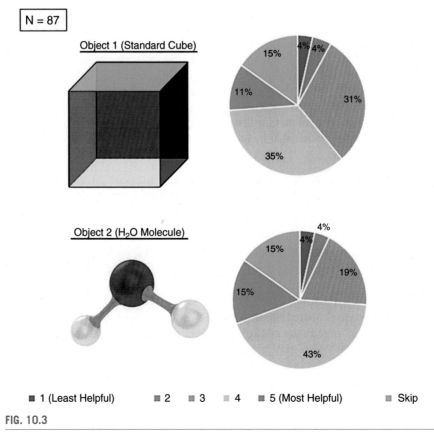

N = 87

Object 1 (Standard Cube)

15% 4% 4% 11% 31% 35%

Object 2 (H_2O Molecule)

4% 4% 15% 19% 15% 43%

■ 1 (Least Helpful) ■ 2 ■ 3 ■ 4 ■ 5 (Most Helpful) ■ Skip

FIG. 10.3

Objects 1 and 2 given to present to them, along with the percents of students who found the mobile application helpful/not helpful in visualizing the corresponding shapes.

Assessment of receptivity

In addition, students were also asked in the postsurvey to rate the extent at which the concept of using the mobile application would help them:

1. Spatially visualize three-dimensional objects that were presented to them.
2. Interconvert two-dimensional and three-dimensional objects.

The intent was to capture the students' perceived usefulness of the mobile application and their receptivity toward it and not to measure for actual SVA (Fig. 10.3).

The results indicate that most students felt that the concept of the mobile application would have been helpful (score of > 3) for them to spatially visualize all four shapes (Fig. 10.4).

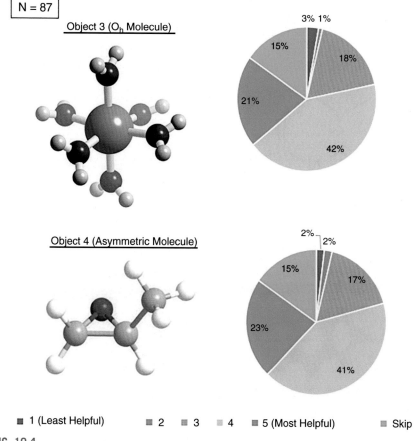

FIG. 10.4

Objects 3 and 4 given to present to them, along with the corresponding percents of students who found the mobile application helpful/not helpful in visualizing the corresponding shapes.

Most students also felt that the concept of the mobile application would be useful for converting two-dimensional to three-dimensional shapes and vice versa. All this indicated that the mobile application was well received (Fig. 10.5).

It was also found that a relatively higher percent of students found the mobile application to be helpful for objects 3 (63%) and 4 (64%) than for shapes 1 (46%) and 2 (57%). This could have been attributed to the fact that the first two objects (standard cube and H_2O molecule) were relatively simpler and that students tended to need more help visualizing the latter two (O_h molecule and asymmetrical molecule). Although the significance of this particular result was not imminent in our main objectives of the trial, it nevertheless still raised potential prospects for further work.

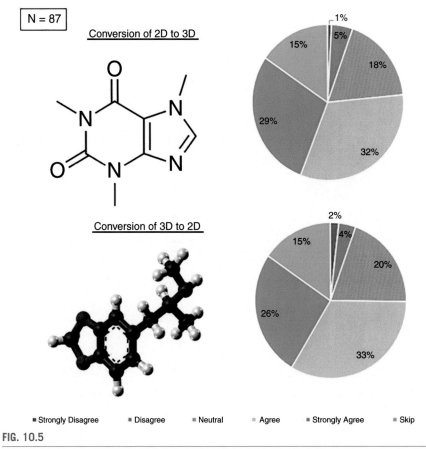

FIG. 10.5

Two-dimensional and three-dimensional objects students were presented with, along with the corresponding percentages of students who agreed/disagreed as to whether the mobile application would help them interconvert the objects into their other form.

Conclusion

Although the idea of using AR in teaching chemistry is not unique, NuPOV could be considered as one of the pioneer applications to promote self-directed learning in students through interacting with the application's interface. Currently, the mobile application prototype has only limited functionality in teaching users to develop better SVA. However, through the trial, it is apparent that our mobile application was well-received by students. With the fundamentals already in place, all that is left is to expand on the functionality of the mobile application by including more complex molecules and reaction mechanisms. Its practical effectiveness has yet to be assessed, and more time is needed to evaluate students' learning with the use of this mobile application.

Acknowledgments

We thank and recognize the contributions from Kevin Christopher Boellaard (figures and conceptualization), Jonah Kailer Aw (analysis), Loh Yi Ping (graphic), and Gan Ju-We (app engineering) for their work in NuPOV. We thank the grant USPC—NUS Grant (2019-01-EDU/USPC-NUS), NUS Department of Chemistry, and support from the Undergraduate Research Opportunities Program (UROPS) Fund and Professor Xavier Coumoul, Vice-President (Culture et Rayonnement) at Université de Paris.

References

[1] Steuer J. Defining virtual reality: dimensions determining telepresence. J Commun 1995;42(4):73–93.

[2] Ryan ML. Narrative as virtual reality: immersion and interactivity in literature and electronic media. Baltimore: Johns Hopkins University Press; 2001.

[3] Bell JT, Fogler HS. The investigation and application of virtual reality as an educational tool. In: Proceedings of the American Society for Engineering Education, Anaheim; 1995.

[4] Wu H-K, Lee SW-Y, Chang H-Y, Liang J-C. Current status, opportunities and challenges of augmented reality in education. Comput Educ 2013;62:41–9.

[5] Yuen SC-Y, Yaoyuneyong G, Johnson E. Augmented reality: an overview and five directions for AR in education. J Educ Technol Dev Exch 2011;4(1).

[6] Grawe DN. Demographics and the demand for higher education. Baltimore: Johns Hopkins University Press; 2018.

[7] Brandenburg R, Mcdonough S, Burke J, White S. Teacher education: innovation, intervention and impact. Singapore: Springer; 2016.

[8] Tatnall A, Okamoto T, Visscher A. Knowledge management for educational innovation, illustrate. New York: Springer; 2006.

[9] Surina L. Dilemmas of the modern educational discourse. Hamburg: Lit Verlag; 2014.

[10] Wright H. Introduction to scientific visualization. Springer Science & Business Media; 2001.

[11] Linsen L, Hagen H, Hamann B, Hege H-C. Visualization in medicine and life sciences II. Berlin: Springer; 2012.

[12] Gilbert JK. Visualization in science education. Springer; 2006.

[13] Cox CT, Poehlmann JS, Ortega C, Lopez JC. Using writing assignments as an intervention to strengthen acid-base skills. J Chem Educ 2018;95(8):1276–83.

[14] Dood AJ, Fields KB, Raker R. Using lexical analysis to predict lewis acid-base model use in responses to an acid-base proton-transfer reaction. J Chem Educ 2018;95(8):1267–75.

[15] Visser T, Maaswinkel T, Coenders F, McKenney S. Writing prompts help improve expression of conceptual understanding in chemistry. J Chem Educ 2018;95(8):1331–5.

[16] Logan K, Mountain L. Writing instruction in chemistry classes: developing prompts and rubrics. J Chem Educ 2018;96(10):1692–700.

[17] Baker RS, Talley L. The relationship of visualisation skills to achievements in freshman chemistry. J Chem Educ 1972;49(11):775.

[18] Lee NY, Tucker-Kellogg G. An accessible, open-source mobile application for macromolecular visualisation using augmented reality. Biochem Mol Biol Educ 2020;48(3):297–303.

[19] Aw JK, Boellaard KC, Tan TK, Yap J, Loh YP, Colasson B, Blanc E, Lam Y, Fung FM. Interacting with three-dimensional molecular structures using an augmented reality mobile app. J Chem Educ 2020;97(10):3877–81.

[20] Kader SN, Ng WB, Tan SWL, Fung FM. Building an interactive immersive virtual reality crime scene for future chemists to learn forensic science chemistry. J Chem Educ 2020;97(9):2651–6.

[21] Fung FM, Choo WY, Ardisara A, Zimmermann CD, Watts S, Koscielniak T, Blanc E, Coumoul X, Dumke R. Applying a virtual reality platform in environmental chemistry education to conduct a field trip to an overseas site. J Chem Educ 2019;96(2):382–6.

[22] Constan M, Ciubotaru Ń. Using virtual reality and augmented reality in education. 1(1); 2008. p. 2–4.

[23] Yang S, Mei B, Yue X. Mobile augmented reality assisted chemical education: insights from elements 4D. J Chem Educ 2018;95:8–10.

[24] Oliver AHJ, Spichkova M, Spencer MJS. Chirality-2: development of a multilevel mobile gaming app to support the teaching of introductory undergraduate-level organic chemistry. J Chem Educ 2018;95(7):1216–20.

[25] Tan ESQ, Soo YJ. Creating apps: a non-IT educator's journey within a higher education landscape. In: Mobile learning in higher education in the Asia-Pacific region. Singapore: Springer; 2017. p. 213–38.

[26] Tasker R, Dalton R. Research into practice: visualization of the molecular world using animations. Chem Educ Res Pract 2006;7(2):141–59.

[27] Fung FM, Lam Y. How COVID-19 disrupted our "flipped" freshman organic chemistry course: insights gained from Singapore. J Chem Educ 2020;97(9):2573–80.

[28] Lee GYP, Lok C, Tan WY, Tan CJY, Tan ESQ. "Are chemistry educational apps useful?"—a quantitative study with three in-house apps. R Soc Chem 2018;19:15–23.

[29] Biirgi HB, Dunitz JD, Shefter E. Geometrical reaction coordinates. II. nucleophilic addition to a carbonyl group. J Am Chem Soc 1973;95(15):5065–7.

[30] Trautz M. Das Gesetz der Reaktionsgeschwindigkeit und der Gleichgewichte in Gasen. Bestätigung der Additivität von Cv-3/2R. Neue Bestimmung der Integrationskonstanten und der Moleküldurchmesser. Z Anorg Allg Chem 1916;96(1):1–28.

[31] Mezirow J. A critical theory of self-directed learning. In: New directions for adult and continuing education; 1985. p. 17–30.

[32] Garrison D. Critical thinking and self-directed learning in adult education. Adult Educ Q 1992;42(3):136–48.

[33] Winer BJ. Statistical principles in experimental design. New York: McGraw-Hill Book Company; 1962.

[34] Liu Y, Kulacki FA. Experimental design. In: Springerbriefs in applied sciences and technology; 2019. p. 49–63.

[35] Becker TE. Potential problems in the statistical control of variables in organisational research: a qualitative analysis with recommendations. Organ Res Methods 2005;8(3):274–89.

[36] Quaiser-Pohl CM, Geiser C. The relationship between computer-game preference, gender, and mental-rotation ability. Personality and individual differences. Personal Individ Differ 2006;40(3):609–19.

[37] Boz Y, Yerdelen-Damar S, Aydemir N, Aydemir M. Investigating the relationships among students' self-efficacy beliefs, their perceptions of classroom learning environment, gender, and chemistry achievement through structural equation modeling. Res Sci Technol Educ 2016;34(3):307–24.

[38] Boujaoude SB. The relationship between students' learning strategies and the change in their misunderstandings during a high school chemistry course. J Res Sci Teach 1992;29(7).

[39] Seery MK. The role of prior knowledge and student aptitude in undergraduate performance in chemistry: a correlation-prediction study. Chem Educ Res Pract 2009;10(3):227–32.

[40] Black AE, Deci EL. The effects of instructors' autonomy support and students' autonomous motivation on learning organic chemistry: a self-determination theory perspective. Sci Educ 2000;84(6):740–56.

[41] Renninger KA, Hidi S, Krapp A. The role of interest in learning and development. New York: Psychology Press; 1992.

[42] Schiefele U. Interest, learning, and motivation. Educ Psychol 1991;26:299–323.

A review of immersive learning technologies featured at EDUCAUSE annual conferences: Evolution since 2016

11

Thierry Koscielniak

Conservatoire National des Arts et Métiers (Le Cnam), DN1 – Direction Nationale des Usages du Numérique, Paris, France

Since 2012, a French delegation of academic people (Professors, CIOs, and CDOs) attended the EDUCAUSE annual conference [1] (hashtags used for these conferences are #EDUxx where xx is the year). The French delegation have produced since 2014 an oral restitution to the French Academic and HigherEdTech community plus a written report [2], in French and translated into English.

This review is a fusion with improvements of a series of articles from the author that began in 2016 in the French delegation's reports: [2]

– 2020 "Immersive learning: grand cru 2020"—to be published in French in March 2020, translated into English in May 2020
– 2019 "Immersive Learning: Massive Feedback in 2019"
– 2018 "Immersive Learning: Promises kept?"
– 2017 "Teaching with virtual reality"
– 2016 "Learning with Virtual Reality"

The evolution of communication types for immersive learning at EDUCAUSE annual conferences:

2016—A preconference seminar/4 sessions/1 poster
2017—One preconference seminar/2 oral sessions/3 posters
2018—One preconference seminar/one XR community group meeting/9 oral sessions/3 posters
2019—Two preconference seminars/one XR community group meeting/8 oral sessions/one corporate presentation/9 posters
2020—One plenary round table/one XR community group meeting/4 oral sessions/one corporate presentation

Technology-Enabled Blended Learning Experiences for Chemistry Education and Outreach
https://doi.org/10.1016/B978-0-12-822879-1.00011-1

#EDU20 Immersive learning: Grand cru 2020

During the 2019 conference, even more experiences were presented than in previous years, and the papers from the 2020 conference were entirely online, in the absence of the two formats—"paid preconference seminars" and "posters." The reduced number of oral presentations is the result of a more selective choice and the feedback is of excellent quality.

Please note, on the EDUCAUSE resources website, there is a page of synthesis documents to start discovering immersive technologies and their pedagogical applications: **Extended Reality** [3].

Abstracts and references for papers in 2019:

Plenary round table

Title: **How Virtual Reality (VR) Makes Diversity Training More Effective** [4]

These were several testimonials on the use of Mursion's VR modules in the selection of candidates and in the training of teachers to recognize and avoid gender bias and to better assist people with disabilities.

These RV simulations use an artificial intelligence engine and it is possible to book a demonstration. Female speakers were enthusiastic about the positive feedback from teachers and students.

Meeting of the XR community group [5]

The theme was unsurprisingly: XR in the world of COVID-19. The meeting took place on the Friday before the conference and unfortunately the author was unable to attend.

The group meets regularly in Webconferences [6] and has a mailing list like all the other community groups of EDUCAUSE (links to register to XR group [7] and to archive [8]).

Oral sessions

Applying an augmented reality mobile app to train visualization skills [9]

The presentation video that was available on demand during the conference is now available on Youtube [10].

"NuPOV" an augmented reality (AR) application for cell phones has been developed at the National University of Singapore to allow students to improve their vision in space when studying three-dimensional models of molecules.

A paper on this work has already been published [11] and the link to the full article available is online [12].

The application was successfully used by the 190 chemistry students. This work allows to validate the interest of using augmented reality in scientific disciplines and to explore evaluation methods using a mobile learning platform that interests and motivates students (Fig. 11.1).

FIG. 11.1

"NuPOV" mobile augmented reality application.

Observation and mentoring with 360-degree video and VR [13]

The aim of the project at the University of Agder (Norway) was to explore how VR and 360-degree video technologies can improve observation practice, tutoring, assessment, and learning outcomes.

One of the highlights of this presentation is that immersive technologies with 360-degree cameras have been used in many different areas and situations:

- Classroom workshop—Department of Nordic Studies and Media
- Teaching in amphitheater for a large group (photo)—Department of Political Science and Management
- Circle session in a crèche (photo)—Teacher Training Department
- Communication between a conductor and his orchestra—Popular Music Department
- Teaching in a gymnasium (photo) and on a ski slope—Physical Education Department
- Therapeutic Conversation Practices—Department of Health Sciences and Nursing

  Student testimonials on the 360-degree video:

- Increases the feeling of presence and intensity by reviewing the video
- Provides more sensations and emotions
- Makes possible the selection of different perspectives and events: Contributes to the sharing of experiences, discoveries, and reflections

Combat: An extended reality experience and other applications [14]

The full description of the presentation is available online [15].

Using Oculus Quest headsets, undergraduate students were able to interact with a 360-degree video documentary about the Battle of Fallujah. The teacher then led a discussion about what it feels like to be in combat. Students expressed that they had a more realistic sense of life in a war zone and felt better equipped to make more informed political decisions. One student commented that "the small size of the enemy combatants' prison cells made me aware of the reality of the situation".

The equipment is part of the new virtual reality program at the Princeton University Library. From the virtual experience of the Battle of Fallujah, informing military intervention and public policy, to practicing public speaking in front of a virtual crowd, the equipment allows students to enhance their learning experiences in a realistic and interactive environment.

Making virtual a reality: Getting immersive technology on campus [16]

Loyola Marymount University has succeeded in integrating immersive technologies into its various colleges in a single semester with a controlled investment [17].

Numerous VR modules have been developed by the school's teachers, facilitating the implementation of modules adapted to the lessons.

A self-service third location has been made available to teachers and students who compare their experiences: *VR Pop-up Lab* [18].

Corporate presentation

Beyond zoom: XR for teaching and research in the COVID-19 era [19].

This HP-sponsored presentation was a synthesis of a conference held on August 4, 2020, organized by Dartmouth College and the University of Pennsylvania. All of the video presentations and their associated documents are available online [20]. The disciplines covered are very diverse (construction, surgery, chemistry, management, geophysics, VR development, electronics, architecture, and third parties) and the feedback is very comprehensive.

#EDU19 immersive learning: Massive feedbacks in 2019

The 2018 review concluded that results are likely to increase over significant numbers of learners. Indeed, the 2019 conference was an opportunity to present even more experiences than in previous years.

The subject of immersive technologies is becoming increasingly important as indicated by the proposal for two preconference seminars that require an additional registration fee.

In the showroom, several manufacturers of VR/AR headsets presented their new products: HP, Lenovo, Microsoft.

Preconference seminars (separate registrations)

Strategic insights into immersive learning: How XR shapes the future [21]?

Ways to help institutions define a strategy on immersive technologies.

Down the rabbit holes: Teaching and learning extended reality technologies [22]

Feedback from the University of Hamilton on creative workshops in Literature using VR. Please note that the slideshow of this paid session is available online [23].

Meeting of the XR working group [24]

An online survey was used to create subworking groups on the two themes: Technologies and Tools/Resources. The group meets regularly in Web conferences.

The slideshow with the agenda of the meeting is available [25].

Oral sessions

XR on campus: Vanguard applications in teaching and learning [26]

The flagship XR presentation of the entire conference. A look back at the second year of the HP/EDUCAUSE "Campus of the Future" partnership that analyzes the use of immersive technologies combined with additive manufacturing (Additive Manufacturing: scanners and 3D printers). Seventeen universities participating in this program in 2018–2019. The results [27] are grouped into the following themes: Objectives, Uses, Integration, Adoption, and Effectiveness.

This presentation has been recorded (video and transcript are subject to a fee except for conference participants).

The slideshow is available online [28].

The progress report of the "Campus of the Future" project is online [27].

Curating a bilateral immersive learning experience: Our France-Singapore story [29]

Professor Xavier Coumoul and the author of this review presented the results of work with their colleagues Dr. Etienne Blanc (Université Paris Descartes) and Dr. Fun Man Fung (National University of Singapore) by students in human toxicology (Paris Descartes University) and environmental toxicology (National University of Singapore). The students created 360-degree resources using Ricoh Theta cameras [30] and the Uptale platform editing studio [31]. A virtual overseas field trip has been also conducted. This work has been published [32] and the full article is available online [33].

Scaling XR teaching and learning: Development, delivery, and assessment strategies [34]

Roundtable of teachers and ISDs from universities: North Carolina State, Penn State, San Diego State, and Sonoma State. Although standards for the design, evaluation, development, and deployment of XR are poorly identified, participants reported progress in the development and refinement of effective practices.

Virtual reality: Engaging, effective, and affordable learning [35]

The Civil Engineer School carried out virtual tours with a 360-degree camera.

The term affordable (affordable and economical) is relative to the American market: the cost of the camera is $3400.

This presentation has been recorded (video and transcript are subject to a fee except for conference participants).

Harsh reality to virtual reality: Getting ahead with immersive tech [36]

Advice from educational engineers at Pennsylvania State university to explore the use of VR and 360-degree shooting. Slideshow (500 Mo!) and poly are available.

Going virtual: VR in higher education [37]

University of Louisiana at Monroe has equipped a room with 28 PCs + Oculus Go headsets to view student work in RV application development.

Anywhere but in the lab: Exploring applications for VR [38]

University of Texas at San Antonio: using VR in Learning Lab; the CIO's point of view; the Labster tool for virtual lab work.

The slideshow is available online [39].

Inspiring innovation: The XReality Center at the New School [40]

Insights about the work of the XReality Center at the New School as a focal point for experimentation, pedagogical innovation, and collaboration. The center's portfolio of curricular implementations, immersive projects, and faculty/students workshops are presented.

Corporate presentation

HP, Yale, and UNL team up on blended reality [41]

Yale and University of Nebraska-Lincoln in partnership with HP on uses of virtual reality in the arts.

Posters

Virtual reality on the bayou [42]

The University of Louisiana Monroe has created a virtual reality center, which includes a classroom with 28 VR headsets, a 3D printer, and two immersive virtual reality rooms.

Enhancing student learning with immersive technologies [43]

Purdue University: Create virtual labs that will be integrated into training pathways to allow off-campus students to participate. Distance education students will have the same experience in the labs as those on campus.

Augmented reality chemistry: A multiyear undergraduate research experience [44]

Georgia Gwinnett College: poster with an interactive presentation of an interdisciplinary and multiyear undergraduate research experience. To develop applications of visualization of molecules using immersive technologies to improve the teaching and learning of chemistry.

Course engagement with immersive visualization [45]

Michigan State University: Integrating immersive technologies into courses to provide students with the opportunity to create visualization experiences.

Visualizing possibilities for virtual reality in education [46]

Learn about the NSF USIgnite grant, through which live streaming VR was used to bring real-world STEM learning to students. An Oculus Go was provided by the presenter to test virtual reality.

The poster is available online [47].

Utilizing VR for enrollment without breaking the bank [48]

Wayne State University uses 360-degree virtual tours of the campus to generate new enrolments.

A toolkit for an immersive VR/AR experience: The Verb Collective [49]

Yale: The Verb Collective is an open set of VR/AR objects designed to help non-programmers (students in the arts and humanities) quickly turn ideas into 3D experiences.

Learning with spatial computing: Virtual worlds, avatars, and 3D collaborative workspaces [50]

VR, AR, and mixed reality will lead to an era of spatial computing where technology blends into the world. Demos of cutting-edge developments in social VR, virtual avatars, and 3D collaborative workspaces. Blueprint and criteria on how to evaluate XR resources and platforms.

Disrupting reality: Virtual and augmented reality in the interior design curriculum [51]

Virtual reality has gone from speculative to ubiquitous. Now in the hands of architects and designers, it has the potential to revolutionize the design process. Uses of virtual reality in interior design and architecture.

#EDU18 Immersive learning: Promises kept?

The evolution of communications on the theme of extended reality (XR eXtended Reality which covers the fields of virtual reality VR, augmented reality AR and mixed reality MR) shows a significant increase in communications in oral sessions in 2018.

Here are the online descriptions and references of all these communications.

Preconference seminar (separate registration required)
Creating immersive storytelling learning experiences in 360-degree video [52]

This full-day session will involve participants in design thinking and prototyping activities to create a short 360° video project. We will examine the challenges of creating a sense of presence, agency, and interactivity. You will gain a new toolset for 360° content creation, design insights, and implementation strategies.

Oral sessions
eXtended reality (XR): The new world of human/machine interaction [53]

eXtended reality (XR) technologies present opportunities to advance the higher education mission and prepare students for a new world of human/machine interaction. In this interactive session, we will explore what is being done today and what is possible in four key areas of XR: use, technology, content development, and gamification.

Step out of your head(set): Better approaches for collaborative learning in virtual environments [54]

The speaker will present her perspective of what VR is as an innovation tool and explore the options we have today to explore virtual spaces to achieve innovation in education and its potential benefits and limitations when it is integrated in a variety of classroom activities. She will discuss her experiences with developing VR systems and applications provide a vision on the future of VR applicability as a force of change education and its potential to become a key innovation technology to improve many aspects of human life.

Applying mixed reality to the classroom of the future [55]

Case Western Reserve University is developing and implementing small and large scale immersive, augmented reality/mixed reality (AR/MR) educational resources, primarily using the Microsoft HoloLens. Case studies in digital alternatives to cadaver-based anatomy in medical school, interactive dance performances, and physics concepts will be presented and explored with attendees.

Holograms in learning: What the real world is telling us [56]

The panel will describe key research questions and methodological design to understand what the data is telling us about mixed reality products as well as about mixed reality in education more broadly.

Virtual reality: Advancing the pedagogical toolkit [57]

In spring 2016, we were approached by a faculty member who wanted to make her Art History 101/102 courses more engaging and improve her students' retention.

We had an opportunity to forge a different path and to try something beyond traditional online tools or standard pedagogical practices. At the time, virtual reality was becoming commercially available, and its transformative potential for the academic community stood out. We will take a deeper look at how VR can shape classrooms from faculty, student, and academic technology perspectives and discuss what many institutions are struggling with: expense and space justification for a technology that is still breaking out of the trough of disillusionment.

Ethics and digital fluency in VR and immersive learning environments [58]

As VR and AR begin to transform the learning environment, we face profound questions regarding digital fluency and ethical issues. What is digital fluency when the virtual and physical worlds are equally real? How will we address the ethical issues and emotional impact of deeply realistic immersive environments?

Developing library strategy for 3D and VR collections [59]

A team led by Virginia Tech (VT) University Libraries, in collaboration with Indiana University (IU) Libraries, and the University of Oklahoma (OU) Libraries will report on an IMLS National Leadership Grant (LG-73-17-0141-17) to organize a series of three national forums on issues of 3D and VR digital collection building.

Virtual holographic simulation: Measuring nursing student outcomes from immersive technology [60]

Emerging holographic VR simulation technology delivers nursing simulation at a fraction of the cost of traditional health care simulation modalities. At San Diego State University, we partnered with Pearson and Microsoft to explore the efficacy of holographic virtual simulation through the HoloLens to increase skills, knowledge, confidence, and motivation to learn.

Mixed reality technology innovation case studies in higher education [61]

Virtual reality, augmented reality, and 3D scanning and printing technologies—collectively called mixed reality (XR) tools—introduce a new set of challenges and opportunities for higher education. We will report on research on opportunities for incorporating XR into the university context, with case studies from Yale, Syracuse, Florida International University, and Hamilton College.

Posters
Bringing the natural sciences online [62]

Arizona State University's College of Liberal Arts and Sciences and ASU Online recently launched two highly popular fully online programs within the natural sciences, biological sciences, and biological chemistry. Come hear about the challenges, innovations, and successes of these programs that use adaptive learning techniques, virtual reality labs, and other technologies.

Bringing virtual reality into higher ed communication [63]

Join us to discuss various ways to integrate virtual reality into higher education digital communication. Learn how Texas A&M is implementing 360 panoramas, cardboard goggles, and high tech VR gear to sell the student experience. It is easier than you think!

The reality of VR: Classrooms for student and community engagement [64]

Virtual reality offers immersive experiences for students, providing active learning experiences not possible in a traditional classroom. The University of Louisiana at Monroe has established a VR classroom with the HTC VIVE kit for use by faculty and students as well as K-12 teachers and students in surrounding parishes.

In 2018, a meeting was also held [65] to establish the *Extended Reality (XR) Community Group* [6] to identify those interested in participating.

In two presentations [56, 60], it is worth noting an abuse of language on the terms hologram or holographic because they refer to Hololens augmented reality glasses created by Microsoft and not to the holographic technology itself.

In an awesome session [54], led by the initial designer of the virtual reality spaces Cave, saw the announcement of a concept of Cave in the box at a reduced cost thanks to mobile equipment. The concept of the Cave corresponds to a room where 3D video projections are made on three out of four walls and the floor/ceiling. A group of people equipped with 3D glasses with plotters can move around virtually while one of the participants has the controllers to interact with the virtual environment.

Cost reduction is achieved by using mobile elements (folding stand walls) and by lowering the price of electronic devices.

A clear trend toward the use of 360-degree shooting technology is noticeable since the cost of equipment is low.

The use of more expensive computer-connected VR headset equipment results in a modeling of training spaces.

What about the answer to the question asked in the title of 2018 paragraph? The promises of the uses of XR (better memorization, total involvement and complete simulations of learning situations) might be proved by increasing the number of experiences. The first results on significant numbers of learners will be available soon.

#EDU17 teaching with virtual reality
Overview and trends

Surprisingly, virtual reality has not been tackled in a massive way in the 2017 edition of the EDUCAUSE conference, unlike artificial intelligence.

Attendees were greeted at the entrance of the show with a Microsoft booth to test Hololens augmented reality glasses on general purpose demos.

The opening plenary lecture by Michio Kaku dealt in part with the subject [66]. According to him, the virtual and augmented realities will mainly change the way we communicate with each other.

A preconference workshop was focused on "Designing Immersive Experiences and Stories in VR/AR that Will Transform Learning [67]," but it was taking place at the time of the campus visits by the French delegation to the 2017 EDUCAUSE annual conference.

A meeting of informal "Meet and Mingle" type meetings was organized on the theme "VR, AR, MR: Immersive Tech" but the author was unfortunately on appointment at the same time.

General questions regarding the use of VR/AR were discussed at the Virtual Worlds Constituent Group meeting [68]. The group was created exactly 10 years ago (January 2008) to discuss the virtual worlds.

In the Exhibit Hall, some stands were demonstrating headsets: Google, Microsoft, Lenovo. Only the xpereal startup presented on its stand a consultant offer in VR/AR/MR (mixed reality)/360-degree videos for education.

During the conference, two sessions of the type "introducing innovation in teaching and learning" treated the subject as well as three posters. It is not much if one considers the number of articles in the specialized press which predict the revolution that will bring the virtual and augmented realities in Education.

Oral sessions

When virtual reality meets the classroom: What happens next [69]?

The VR initiative is part of a global one called MOSAIC from Indiana University on Active Learning. Many innovative Learning Spaces are presented and linked to an Advanced Visualization Lab that integrates all 3D imaging technologies is simulation.

This led to the need to create Reality Labs to test AR and VR.

Reality Labs rooms are comprised of powerful computers for VR, high-definition displays and Oculus-style headsets.

Many VR applications are tested in various fields: art, interior decoration, architectural archeology, anatomy, astronomy, music, etc.

The next step of Reality Labs is the settlement in spring 2018 of a Creative Lab named Idea Garden. There will be tested in portable devices (iBackpack), augmented reality glasses and 360-degree cameras associated with 3D scanners and 3D printers.

Making virtual reality a reality: Applications of augmented/virtual reality [70]

The presentation took place in three steps:

- Description of the steps for creating a VR Lab; types of equipment and costs.
- The Kinber network of the State of Pennsylvania to fund 360-degree camera uses
- Presentation of many experiences of VR uses.

This session was extremely useful for someone wanting to discover what VR is and its applications. The three slide shows are available (see link in the note) and exhaustive.

Posters

Immersive learning with 360 video [71]

This poster shows the use of 360-degree videos (video-spheres) to show how machine tools work for engineering students at Penn State University.

The videos are ported over to Adobe Premiere Pro which is provided free to all Penn State students. But only editing functions are used.

The next step could be to use a video-spheres editing tool to make them interactive. For example, the Uptale [31] online sofware, from FrenchTech companies.

The upcoming opening of an Immersive Experiences Lab at PennState is announced; resources are already online.

The use of virtual reality technologies in architecture instruction and critiques [72]

The authors propose to provide architecture students with simulation platforms to perform critical reports and tests.

Virtual/mixed/augmented reality overview fall, 2017 [73]

Poster author Susan Molnar described the different techniques and materials for using VR/AR. His core business is to be an artist and educator [74].

To complete your discovery of virtual reality, you can read of the reference books (five volumes) "Treaty of Virtual Reality" [75] coordinated by Philippe Fuchs, professor at the Ecole des Mines ParisTech.

Meanwhile in France, there was also an initiative to create an Immersive Learning Lab (i2L) in Paris by Nicolas Dupain to bring together French actors around an innovative ecosystem. A kick-off meeting has been at le Cnam in February 2018. All information on i2L is on its website (in French) [76].

#EDU16 Learning with virtual reality

The term Virtual Reality covers the technological devices that allow "simulating an environment with which the user can interact" (French version of Wikipedia article [77]). The correct term to use according to this Wikipedia article would rather be "Realistic Virtuality."

This review is mainly based on the "Virtual Reality (VR) in the Classroom" [78] session presented on October 26 at EDUCAUSE 2016 by Andrew Lloyd Goodman and Kelly Egan of Brown University.

Other sessions have addressed this topic, but the difficulty at the EDUCAUSE conference is the number of sessions in parallel (about 20) and the resulting Cornelian choice. Here are the references of the other sessions:

Cup of gold: Designing and developing a virtual reality learning space [79]

This presentation will commence with a demonstration of Cup of Gold, a web-based virtual reality game designed to teach basic information literacy skills and introduce incoming students to the scale of academic libraries. Following the demo, we will present a summary of aggregated user data, highlight key design elements, and explain how these elements were incorporated into the development process.

Virtual Worlds [80]
Create virtual learning environments to improve the student experience [81]

Today digital transformation is opening a new world of learning opportunities. This round table discussion with Georgia State University and Mercer University will explore digital in higher education: how students learn with mobile devices anywhere, anytime; how instructors engage in classes and online communities through interactive video and virtual labs; and how online courses expand reach and complement traditional instruction, creating global classrooms.

A self-managed multilingual virtual classroom [82]

Learn about methodologies and tools for promoting multilingual virtual classrooms where the environment is able to adapt to the users automatically without requiring any further effort on their part.

A paid seminar preconference was also dedicated to this topic: "Virtual Reality and the Future of Learning" [83]

This experiential seminar will explore the learning opportunities afforded through the development of virtual reality devices, platforms, and experiences. The presenters will review projects that can be implemented in the classroom, makerspaces, innovation centers, and VR labs. Virtual reality offers immersive interactive experiences that create compelling learning resources and environments. These developments require innovation in pedagogy, learning space design, and institutional culture.

An immersive VR research project

First, the Visualization Research Laboratory (VRL [84]) at Brown University and the work of Professor David H. Laidlaw on the subject were presented. The goal of the project is to provide a graphical environment for researchers to visualize their data. The project is called YURT a recursive acronym: "YURT Ultimate Reality Theater." It refers to the yurt-like shape of the visualization space: a conical roof and cylindrical walls on which images are projected in which the user is immersed. The graphic resolution of the images is pushed to the maximum to give the user the most realistic impression possible.

From a technical point of view, the YURT project uses 20 computers connected to 69 3D video projectors with 145 mirrors. The total resolution is more than 100 million pixels. The user is immersed in the three-dimensional images floating around him. The potential of this technology for teaching is enormous. Students have been immersed in protein structures to help them understand how they fold.

The artistic dimension is not forgotten because the laboratory opens its doors to Professor John Cayley who realizes "space poems" with his students and to Adam Blumenthal, artist in residence in the laboratory. A research work in archeology by Professor Laurel Bestock is in progress.

The conclusion of Professor Laidlaw is that in 15–20 years, the screens will have disappeared and that we will be surrounded by our friends in the form of 3D avatars for uses that are still to be imagined through science fiction movies. (This last sentence written in 2016 is full reality in 2020 with "Ready Player One" Movie.)

The three examples of uses of the YURT immersion space were then detailed.

Writing in immersive VR [85]
John Cayley, professor of literature, uses YURT to Brown so that his students can create 3D texts as part of his teaching of "digital language in electronic poetic writing." The user is immersed in an artistic 3D representation of the texts. John Cayley thinks we are emerging from the "winter of virtual reality": a real artificial three-dimensional space of writing is born.

Recreate an historical event [86]
Adam Blumenthal is an artist whose project is to describe the "Gaspee Affair [87]" using virtual reality glasses (type Oculus Rift or HTC Vive) and the YURT. He appreciates the ability to move around in environments that are normally impossible to access, for example, by returning to the past. The historical event studied was one of the triggers of the declaration of independence of the United States. The artistic work produced [88] will allow a machine effect to go back in time by reliving tragic events of boat chases, battles, and explosions while presenting all the background documents, paintings, maps and objects of the boat Gaspée. Interactivity will be omnipresent to allow a nonlinear path to the user.

One obstacle to be overcome is the discomfort caused by wearing large spectacles attached to a cable. The user is limited in his walk by this "connected leash." Unlike the YURT, the use of glasses is a solo experience. But using Google Cardboards will allow hundreds of students to immerse themselves at the same time.

The creation tools used in augmented reality are the same as those that produce video games and they are "free."

Visualize curious excavations in the YURT [89]
Laurel Bestock, professor of Egyptology and Archeology, uses the YURT to visualize the excavation spaces with his students. Archeology is intrinsically three-dimensional because it studies the relationships between objects and their environment and their meanings within an era and a culture. Modeling a search in northern Sudan, part of ancient Nubia. On-site presence is possible from 2 to 3 weeks per year. It is essential to create numerical modeling of an immense fortress implanted over hundreds of meters and composed of more than three million bricks. Initially, the modeling was not

intended to be used with virtual reality but rather to create mappings. Viewing in the YURT data from a house took only 5 min. All the construction history then appeared clearly to the archaeologists as if they were moving in the field of excavations. New prospects for virtual excavations have arisen because if archeology is destructive, 3D modeling makes it possible to reconstruct virtually what has been excavated, or even to excavate the same site in different ways.

From VR to AR

Virtual reality at Brown allows for innovative scientific and artistic uses in education and research. Other institutions and companies are of course present in this field. As a reminder, the French delegation at EDUCAUSE 2013 visited the University of California at San Diego (UCSD) Wave [90], laboratory, another virtual reality immersion facility.

Beyond the sometimes disorienting effect for the senses of being immersed for too long in a virtual world through a headset, the author was able to experiment the use of augmented reality glasses. This time, it is a question of transparent glasses with a translucent projection area that superimposes digital objects with the user's environment. The Google Glasses are the precursor model with a very small area located in the periphery of a single eye. They have the advantage of being wireless, but their marketing was stopped in January 2015. The MetaGlasses (Metavision Company) were used in the challenges of "La Nuit du Numérique 2015" organized by the author at Paris Descartes University, in Paris, France. The use of the SDK (Software Development Kit) by the computer science students allowed them to quickly create prototypes with the support of a Meta engineer who came directly from Silicon Valley. Unfortunately the presence of the cable still limits the movements in this prototype. Finally, the long-awaited Hololens [91] from Microsoft keep their promises as a visualization in a space mapped in real time by the embedded 3D technology. The absence of a cable is an asset that has its downside: poor autonomy.

The 3D glasses, virtual or augmented, prefigure new pedagogical interfaces that will be functional in the medium term. Immersive spaces are still at a price that makes them difficult to generalize in the teaching places.

References

[1] EDUCAUSE. EDUCAUSE annual conference. [Online]. Available: https://events.educause.edu/annual-conference; 2021. [Accessed March 2021].

[2] Augeri J, Flory L, Koscielniak T, Urbero B, Verez D, et al. [Online]. Available: http://tinyurl.com/delegation-Fr-EDUCAUSE; 2021. [Accessed March 2021].

[3] EDUCAUSE. Extended reality (XR). [Online]. Available: https://library.educause.edu/topics/emerging-technologies/extended-reality-xr; 2021. [Accessed March 2021].

[4] Bondie R, Brooks L, Ruffin M, Straub C. How virtual reality makes diversity training more effective. [Online]. Available: https://events.educause.edu/annual-conference/2020/agenda/featured-session-2; 27 October 2020. [Accessed March 2021].

[5] McCreary B, Sprecher A. Extended reality (XR) community group session. XR Community Group; 23 October 2020. [Online]. Available: https://events.educause.edu/annual-conference/2020/agenda/extended-reality-xr-community-group-session. [Accessed March 2021].

[6] EDUCAUSE. XR community group. [Online]. Available: https://www.educause.edu/community/xr-extended-reality-community-group; 2021. [Accessed March 2021].

[7] EDUCAUSE. XR@LISTSERV.EDUCAUSE.EDU. XR Community Group; 2021. [Online]. Available: http://listserv.educause.edu/scripts/wa.exe?SUBED1=xr&A=1. [Accessed March 2021].

[8] EDUCAUSE. Archives for XR@LISTSERV.EDUCAUSE.EDU. [Online]. Available: http://listserv.educause.edu/scripts/wa.exe?A0=xr; 2021. [Accessed March 2021].

[9] Fung FM, Yap J. Applying an augmented reality mobile app to train visualization skills. [Online]. Available: https://events.educause.edu/annual-conference/2020/agenda/applying-an-augmented-reality-mobile-app-to-train-visualization-skills; October 2020. [Accessed March 2021].

[10] Fung FM, Director. Augmented reality mobile app to train visualization skills; 2020 [Film]. https://www.youtube.com/watch?v=7UUA0xaTegM.

[11] Aw JK, Boellaard KC, Tan TK, Yap J, Loh YP, Colasson B, Blanc É, Lam Y, Fung FM. Interacting with three-dimensional molecular structures using an augmented reality mobile app. J Chem Educ 2020;97(10):3877–81.

[12] Aw JK, Boellaard KC, Tan TK, Yap J, Loh YP, Colasson B, Blanc É, Lam Y, Fung FM. Interacting with three-dimensional molecular structures using an augmented reality mobile app. [Online]. Available: https://pubs.acs.org/doi/10.1021/acs.jchemed.0c00387. [Accessed March 2021].

[13] Skaar SE. Observation and mentoring with 360° video and VR. [Online]. Available: https://events.educause.edu/annual-conference/2020/agenda/observation-and-mentoring-with-360-video-and-vr; 27 October 2020. [Accessed March 2021].

[14] Porter S. Combat: an extended reality experience and other applications. [Online]. Available: https://events.educause.edu/annual-conference/2020/agenda/combat-an-extended-reality-experience-and-other-applications; October 2020. [Accessed March 2021].

[15] Ramírez S. Virtual reality programming at PUL offers princeton opportunities to learn in an immersive extended reality. [Online]. Available: https://library.princeton.edu/news/general/2020-02-06/virtual-reality-programming-pul-offers-princeton-opportunities-learn. [Accessed March 2021].

[16] Henline J, Schwartz J. Making virtual a reality: getting immersive technology on campus. [Online]. Available: https://events.educause.edu/annual-conference/2020/agenda/making-virtual-a-reality-getting-immersive-technology-on-campus; October 2020. [Accessed March 2021].

[17] Loyola Marymount University. Immersive technology. [Online]. Available: https://its.lmu.edu/whatwedo/instructionaltechnology/immersivetechnology/; 2021. [Accessed March 2021].

[18] Loyola Marymount University. LMU VR Pop-Up Lab. [Online]. Available: https://its.lmu.edu/vr/; 2019. [Accessed March 2021].

[19] Bell J, Castro D. Beyond zoom: XR for teaching and research in the COVID-19 era. [Online]. Available: https://events.educause.edu/annual-conference/2020/agenda/beyond-zoom-xr-for-teaching-and-research-in-the-covid19-era; 27 October 2020. [Accessed March 2021].

[20] Dartmouth College and University of Pennsylvania. Beyond Zoom: XR for teaching and research in the COVID-19 era program. [Online]. Available: https://dartgo.org/covid-xr-conf; 07 August 2020. [Accessed March 2021].

[21] Craig E, Georgiva M. Strategic insights into immersive learning: how XR shapes the future. [Online]. Available: https://events.educause.edu/annual-conference/2019/agenda/strategic-insights-into-immersive-learning-how-xr-shapes-the-future-separate-registration-is-require; 14 October 2019. [Accessed March 2021].

[22] Higgins D, Salzman B, Serrano N. Down the rabbit holes: teaching and learning extended reality technologies. [Online]. Available: https://events.educause.edu/annual-conference/2019/agenda/down-the-rabbit-holes-teaching-and-learning-extended-reality-technologies-separate-registration-is-r; 14 October 2019. [Accessed March 2021].

[23] Higgins D, Salzman B, Serrano N. Session powerpoint down the rabbit holes teaching learning extended reality technologies. [Online]. Available: https://events.educause.edu/HubbEventResources/E19/SEM08A/SEM08A%20-%20Down_the_Rabbit_Holes__Teaching___Learning_Extended_Reality_Technologies.pdf; 14 October 2019. [Accessed March 2021].

[24] McCreary B, Sprecher A. Extended reality (XR) community group session. [Online]. Available: https://events.educause.edu/annual-conference/2019/agenda/extended-reality-xr-community-group-session-open-to-all; 17 October 2019. [Accessed March 2021].

[25] McCreary B, Sprecher A. Slides extended reality (XR) community group session. [Online]. Available: https://events.educause.edu/HubbEventResources/E19/DISC35/DISC35%20-%20eXtended_Reality__XR__Community_Group_Meeting_2019-10_FINAL.pptx; 17 October 2019. [Accessed March 2021].

[26] Brown M, Hibbert M, Pomerantz J, Thompson M. XR on campus: vanguard applications in teaching and learning. [Online]. Available: https://events.educause.edu/annual-conference/2019/agenda/xr-on-campus-vanguard-applications-in-teaching-and-learning; 15 October 2019. [Accessed March 2021].

[27] Pomerantz J. XR for teaching and learning. [Online]. Available: https://library.educause.edu/resources/2019/10/xr-for-teaching-and-learning; 10 October 2019. [Accessed March 2021].

[28] Pomerantz J. XR on campus pomerantzs slides. [Online]. Available: https://events.educause.edu/HubbEventResources/E19/SESS191C/SESS191C%20-%20XRonCampus_Pomerantz_Slides.pdf; 11 December 2019. [Accessed March 2021].

[29] Blanc E, Xavier C, Fung FM, Koscielniak T. Curating a bilateral immersive learning experience: our France-Singapore story. [Online]. Available: https://events.educause.edu/annual-conference/2019/agenda/curating-a-bilateral-immersive-learning-experience-our-francesingapore-story; 17 October 2019. [Accessed March 2021].

[30] Ricoh Company Limited. THETA. [Online]. Available: https://theta360.com/en/; 2021. [Accessed March 2021].

[31] UPTALE. Enterprise immersive learning solution. [Online]. Available: https://www.uptale.io/en/home/. [Accessed March 2021].

[32] Fung FM, Choo WY, Ardisara A, Zimmermann CD, Watts S, Koscielniak T, Blanc E, Coumoul X, Dumke R. Applying a virtual reality platform in environmental chemistry education to conduct a field trip to an overseas site. J Chem Educ 2019;96(2):382–6.

[33] Fung FM, Choo WY, Ardisara A, Zimmermann CD, Watts S, Koscielniak T, Blanc E, Coumoul X, Dumke R. Applying a virtual reality platform in environmental chemistry education to conduct a field trip to an overseas site. [Online]. Available: https://pubs.acs.org/doi/full/10.1021/acs.jchemed.8b00728; 25 January 2019. [Accessed March 2021].

[34] Bowen K, Frazee J, Hauze S, Kassis S, Woodbury D. Scaling XR teaching and learning: development, delivery, and assessment strategies. [Online]. Available: https://events.educause.edu/annual-conference/2019/agenda/scaling-xr-teaching-and-learning-development-delivery-and-assessment-strategies; 16 October 2019. [Accessed March 2021].

[35] Fuller T, Kappel R. Virtual reality: engaging, effective, and affordable learning. [Online]. Available: https://events.educause.edu/annual-conference/2019/agenda/virtual-reality-engaging-effective-and-affordable-learning; 15 October 2019. [Accessed March 2021].

[36] Byrd J, Ings H, Lawrence C. Harsh reality to virtual reality: getting ahead with immersive tech. [Online]. Available: https://events.educause.edu/annual-conference/2019/agenda/harsh-reality-to-virtual-reality-getting-ahead-with-immersive-tech; 19 October 2019. [Accessed March 2021].

[37] Hoover T. Going virtual: VR in higher education. [Online]. Available: https://events.educause.edu/annual-conference/2019/agenda/going-virtual-vr-in-higher-education; 15 October 2019. [Accessed March 2021].

[38] Abel J, Kenon V, Ketchum K, Nash T. Anywhere but in the lab: exploring applications for VR. [Online]. Available: https://events.educause.edu/annual-conference/2019/agenda/anywhere-but-in-the-lab-exploring-applications-for-vr; 16 October 2019. [Accessed March 2021].

[39] Abel J, Kenon V, Ketchum K, Nash T. AnywhereButInTheLabSlidesppt. [Online]. Available: https://events.educause.edu/HubbEventResources/E19/SESS120/SESS120%20-%20Educause_2019_10_10_19.pptx; 26 November 2019. [Accessed March 2021].

[40] Georgieva M. Inspiring innovation: the XReality center at the new school. [Online]. Available: https://events.educause.edu/annual-conference/2019/agenda/inspiring-innovation-the-xreality-center-at-the-new-school; 17 October 2019. [Accessed March 2021].

[41] Berry J, Elliott M, Rode R, Webb J. HP, yale, and UNL team up on blended reality. [Online]. Available: https://events.educause.edu/annual-conference/2019/agenda/hp-yale-and-unl-team-up-on-blended-reality; 15 October 2019. [Accessed March 2021].

[42] Hoover T. Virtual reality on the bayou. [Online]. Available: https://events.educause.edu/annual-conference/2019/agenda/virtual-reality-on-the-bayou; 15 October 2019. [Accessed March 2021].

[43] Hopkins A, Takahashi G. Enhancing student learning with immersive technologies. [Online]. Available: https://events.educause.edu/annual-conference/2019/agenda/enhancing-student-learning-with-immersive-technologies; 15 October 2019. [Accessed March 2021].

[44] Behmke D, Brannock E, Lutz B. AR chemistry: a multiyear undergrad research experience. [Online]. Available: https://events.educause.edu/annual-conference/2019/agenda/ar-chemistry-a-multiyear-undergraduate-research-experience; 16 March 2019. [Accessed March 2021].

[45] O'Neill T. Course engagement with immersive visualization. [Online]. Available: https://events.educause.edu/annual-conference/2019/agenda/course-engagement-with-immersive-visualization; 16 October 2019. [Accessed March 2021].

[46] Davis R. Visualizing possibilities for virtual reality in education. [Online]. Available: https://events.educause.edu/annual-conference/2019/agenda/visualizing-possibilities-for-virtual-reality-in-education; 16 October 2019. [Accessed March 2021].

[47] Davis R. Poster visualizing possibilities for VR. [Online]. Available: https://events.educause.edu/HubbEventResources/E19/PS160a/PS160a%20-%20EDUCAUSErdavis.pdf; 16 October 2019. [Accessed March 2021].

[48] Crabtre M, Hubbard D, Medley D. Utilizing VR for enrollment without breaking the bank. [Online]. Available: https://events.educause.edu/annual-conference/2019/agenda/utilizing-vr-for-enrollment-without-breaking-the-bank; 15 October 2019. [Accessed March 2021].

[49] Rode R. A toolkit for an immersive VR/AR experience: the verb collective. [Online]. Available: https://events.educause.edu/annual-conference/2019/agenda/a-toolkit-for-an-immersive-vrar-experience- -the-verb-collective; 16 October 2019. [Accessed March 2021].

[50] Craig E, Georgieva M. Learning with spatial computing: virtual worlds, avatars, and 3D collaborative workspaces. [Online]. Available: https://events.educause.edu/annual-conference/2019/agenda/learning-with-spatial-computing-virtual-worlds-avatars-and-3d-collaborative-workspaces; 15 October 2019. [Accessed March 2021].

[51] Duran R, Siddiqi M. Disrupting reality: virtual and augmented reality in the interior design curriculum. [Online]. Available: https://events.educause.edu/annual-conference/2019/agenda/disrupting-reality-how-vr-is-changing-interior-design; 16 October 2019. [Accessed March 2021].

[52] Craig E, Georgieva M. Creating immersive storytelling learning experiences in 360° video. [Online]. Available: https://events.educause.edu/annual-conference/2018/agenda/creating-immersive-storytelling-learning-experiences-in-360-video-separate-registration-is-required; 30 October 2018. [Accessed March 2021].

[53] Diener S, McCreary B, Sprecher A. eXtended Reality (XR): the new world of human/machine interaction. [Online]. Available: https://events.educause.edu/annual-conference/2018/agenda/extended-reality-xr-the-new-world-of-humanmachine-interaction; 31 October 2018. [Accessed March 2021].

[54] Cruz-Neira C. Step out of your head(set): better approaches for collaborative learning in virtual environments. [Online]. Available: https://events.educause.edu/annual-conference/2018/agenda/step-out-of-your-headset-better-approaches-for-collaborative-learning-in-virtual-environments; 31 October 2018. [Accessed March 2021].

[55] Griswold M, Henninger E, Shick S. Applying mixed reality to the classroom of the future. [Online]. Available: https://events.educause.edu/annual-conference/2018/agenda/applying-mixed-reality-to-the-classroom-of-the-future; 31 October 2018. [Accessed March 2021].

[56] Christian M, Frith G, Geoghan D. Holograms in learning: what the real world is telling us. [Online]. Available: https://events.educause.edu/annual-conference/2018/agenda/holograms-in-learning-what-the-real-world-is-telling-us; 01 November 2018. [Accessed March 2021].

[57] Dobbin-Bennett T, Foster S, Ward L. Virtual reality: advancing the pedagogical toolkit. [Online]. Available: https://events.educause.edu/annual-conference/2018/agenda/virtual-reality-advancing-the-pedagogical-toolkit; 01 November 2018. [Accessed March 2021].

[58] Craig E, Georgieva M. Ethics and digital fluency in VR and immersive learning environments. [Online]. Available: https://events.educause.edu/annual-conference/2018/agenda/ethics-and-digital-fluency-in-vr-and-immersive-learning-environments; 02 November 2018. [Accessed March 2021].

[59] Cook M, Griffin J, McDonald R. Developing library strategy for 3D and VR collections. [Online]. Available: https://events.educause.edu/annual-conference/2018/agenda/developing-library-strategy-for-3d-and-virtual-reality-collections; 02 November 2018. [Accessed March 2021].

[60] Frazee J, Hauze S. Virtual holographic simulation: measuring nursing student outcomes from immersive technology. [Online]. Available: https://events.educause.edu/annual-conference/2018/agenda/virtual-holographic-simulation-measuring-nursing-student-outcomes-from-immersive-technology; 02 November 2018. [Accessed March 2021].

[61] Brooks DC, Rode R, Salzman B, Stuart J, Webb J. Mixed reality technology innovation case studies in higher ed. [Online]. Available: https://events.educause.edu/annual-conference/2018/agenda/mixed-reality-technology-innovation-case-studies-in-higher-education; 01 November 2018. [Accessed March 2021].

[62] Austin A, Garcia G, Pate A, Polk L, Robinson M. Bringing the natural sciences online. [Online]. Available: https://events.educause.edu/annual-conference/2018/agenda/bringing-the-natural-sciences-online; 31 October 2018. [Accessed March 2021].

[63] Green M, Hushek C. Bringing virtual reality into higher ed communication. [Online]. Available: https://events.educause.edu/annual-conference/2018/agenda/bringing-virtual-reality-into-higher-ed-communication; 31 October 2018. [Accessed March 2021].

[64] Hoover T. The reality of VR: classrooms for student and community engagement. [Online]. Available: https://events.educause.edu/annual-conference/2018/agenda/the-reality-of-vr-classrooms-for-student-and-community-engagement; 01 November 2018. [Accessed March 2021].

[65] McCreary B, Sprecher A. Extended reality (XR) community group session. [Online]. Available: https://events.educause.edu/annual-conference/2018/agenda/extended-reality-xr-community-group-session-open-to-all; 31 October 2018. [Accessed March 2021].

[66] Kaku M. The next 20 years in education and technology. [Online]. Available: https://events.educause.edu/annual-conference/2017/agenda/the-next-20-years-in-education-and-technology; 01 November 2017. [Accessed March 2021].

[67] Craig E, Georgieva M. Designing immersive experiences and stories in VR/AR that will transform learning. [Online]. Available: https://events.educause.edu/annual-conference/2017/agenda/sem02adesigning-immersive-experiences-and-stories-in-vrar-that-will-transform-learning-separate-registration-is-required; 31 October 2017. [Accessed March 2021].

[68] Diener S. Virtual worlds constituent group meeting. [Online]. Available: https://events.educause.edu/annual-conference/2017/agenda/virtual-worlds-constituent-group-meeting; 02 November 2017. [Accessed March 2021].

[69] Boyles M, Johnston J. When virtual reality meets the classroom: what happens next? [Online]. Available: https://events.educause.edu/annual-conference/2017/agenda/when-virtual-reality-meets-the-classroom--what-happens-next; 03 November 2017. [Accessed March 2021].

[70] Belloff T, Fineman B, Oxenford J, Pellegrini M. Making virtual reality a reality: applications of augmented/virtual reality. [Online]. Available: https://events.educause.edu/annual-conference/2017/agenda/making-virtual-reality-a-reality-applications-of-augmentedvirtual-reality; 01 November 2017. [Accessed March 2021].

[71] Wetzel R. Immersive learning with 360 video. [Online]. Available: https://events.educause.edu/annual-conference/2017/agenda/immersive-learning-with-360-video; 02 November 2017. [Accessed March 2021].

[72] Finnell M, Hart A, Schipps D, Tanz K. The use of virtual reality technologies in architecture instruction and critiques. [Online]. Available: https://events.educause.edu/annual-conference/2017/agenda/the-use-of-virtual-reality-technologies-in-architecture-instruction-and-critiques; 01 November 2017. [Accessed March 2021].

[73] Molnar S. Virtual/mixed/augmented reality overview fall 2017. [Online]. Available: https://events.educause.edu/annual-conference/2017/agenda/virtualmixedaugmented-reality-overview-fall-2017; 02 November 2017. [Accessed March 2021].

[74] Molnar S. SuMo design. [Online]. Available: http://www.susanmolnar.com/; 2021. [Accessed March 2021].

[75] Fuchs P. Le traité de la réalité virtuelle. [Online]. Available: https://www.pressesdes-mines.com/?s=realit%C3%A9+virtuelle&post_type=product; 2021. [Accessed March 2021].

[76] France Immersive Learning. Immersive learning lab. [Online]. Available: https://fran-ceimmersivelearning.fr/immersivelearninglab/; 2021. [Accessed March 2021].

[77] WIKIPEDIA. Réalité virtuelle. [Online]. Available: https://fr.wikipedia.org/wiki/R%C3%A9alit%C3%A9_virtuelle; 10 March 2021. [Accessed March 2021].

[78] Egan K, Goodman AL. Virtual reality in the classroom. [Online]. Available: https://events.educause.edu/annual-conference/2016/agenda/virtual-reality-in-the-classroom; 26 October 2016. [Accessed March 2021].

[79] Rechnitz A. Cup of gold: designing and developing a virtual reality learning space. [Online]. Available: https://events.educause.edu/annual-conference/2016/agenda/cup-of-gold-designing-and-developing-a-virtual-reality-learning-space; 28 October 2016. [Accessed March 2021].

[80] Diener S. Virtual worlds. [Online]. Available: https://events.educause.edu/annual-confer-ence/2016/agenda/virtual-worlds; 26 October 2016. [Accessed March 2021].

[81] Adkins T, Milam S, Patton R. Create virtual learning environments to improve the student experience. [Online]. Available: https://events.educause.edu/annual-conference/2016/agenda/create-virtual-learning-environments-to-improve-the-student-experience; 26 October 2016. [Accessed March 2021].

[82] Duran Cals J. A self-managed multilingual virtual classroom. [Online]. Available: https://events.educause.edu/annual-conference/2016/agenda/a-selfmanaged-multilingual-vir-tual-classroom; 26 October 2016. [Accessed March 2021].

[83] Craig E, Georgieva M. Virtual reality and the future of learning. [Online]. Available: https://events.educause.edu/annual-conference/2016/agenda/sem01avirtual-reality-and-the-future-of-learning-separate-registration-is-required; 25 October 2016. [Accessed March 2021].

[84] Brown University. Visualization research lab. [Online]. Available: http://vis.cs.brown.edu; 30 April 2018. [Accessed March 2021].

[85] Goodman and Andrew, Directors. John Cayley—The CAVE; 2016 [Film]. https://youtu.be/wqwQTzP3T0s?list=PLipxZuLedV-NdA59QiQ8d36AMfmvIPiAa.

[86] Goodman A, Director. Adam Blumenthal—HTC Vive; 2016 [Film]. https://youtu.be/2Zk85mo6NVM?list=PLipxZuLedV-NdA59QiQ8d36AMfmvIPiAa.

[87] WIKIPEDIA. Gaspee affair. [Online]. Available: https://en.wikipedia.org/wiki/Gaspee_Affair; 12 March 2021. [Accessed March 2021].

[88] CURIOUS SENSE. WORK, POSTS & PRESS image 8 Jun Brown University Names Adam Blumenthal virtual reality artist-in-residence. [Online]. Available: https://curi-oussense.com/brown-university-names-adam-blumenthal-virtual-reality-artist-in-resi-dence/; 2021. [Accessed March 2021].

[89] Goodman A, Director. Laurel Bestock—The YURT; 2016 [Film]. https://youtu.be/3kMMSq4Xuc8?list=PLipxZuLedV-NdA59QiQ8d36AMfmvIPiAa.

[90] University of California—San Diego. Visualization systems. [Online]. Available: http://chei.ucsd.edu/wave/; 2021. [Accessed March 2021].

[91] MICROSOFT. HoloLens (1st gen) hardware. [Online]. Available: https://docs.microsoft.com/en-us/hololens/hololens1-hardware; 16 September 2019. [Accessed March 2021].

Index

Note: Page numbers followed by *f* indicate figures and *t* indicate tables.

Printed in the United States
by Baker & Taylor Publisher Services